Electrical Engineering Practice Problems

for the Power, Electrical and Electronics, and Computer PE Exams

Eighth Edition

John A. Camara, PE

Professional Publications, Inc. • Belmont, CA

How to Locate and Report Errata for This Book

At Professional Publications, we do our best to bring you error-free books. But when errors do occur, we want to make sure you can view corrections and report any potential errors you find, so the errors cause as little confusion as possible.

A current list of known errata and other updates for this book is available on the PPI website at **www.ppi2pass.com/errata**. We update the errata page as often as necessary, so check in regularly. You will also find instructions for submitting suspected errata. We are grateful to every reader who takes the time to help us improve the quality of our books by pointing out an error.

National Electrical Code® and NEC® are registered trademarks of the National Fire Protection Association, Inc., Quincy, MA 02169.

National Electrical Safety Code® and NESC® are registered trademarks of the Institute of Electrical and Electronics Engineers, Inc.

ELECTRICAL ENGINEERING PRACTICE PROBLEMS
Eighth Edition

Current printing of this edition: 1

Printing History

edition number	printing number	update
6	4	Minor corrections.
7	1	New edition. Code updates.
8	1	New edition. New practice problems. Code updates.

Printed in the United States of America

PPI
1250 Fifth Avenue, Belmont, CA 94002
(650) 593-9119
www.ppi2pass.com

ISBN-13: 978-1-59126-112-4
ISBN-10: 1-59126-112-0

Library of Congress Control Number: 2008937992

Topics

Mathematics

Theory

Field Theory

Circuit Theory

Power—Generation

Power—Transmission

Power—Machinery

Power—Special Apps

Measurement & Instrumentation

Electronics

Computers

Communications

Biomedical Systems

Control Systems

Electrical Materials

Codes and Standards

Professional

Where do I find help
solving these Practice Problems?

Electrical Engineering Practice Problems for the Power, Electrical and Electronics, and Computer PE Exams presents complete, step-by-step solutions for 460 problems covering topics on the Power, Electrical and Electronics, and Computer PE exams. This book is a companion to the *Electrical Engineering Reference Manual*; the subjects covered in the two books correspond chapter for chapter, and the solutions in this book often cite the *Reference Manual*. You can find all the background information you need to solve these problems—including equations, figures, charts, and tables of data—in the *Electrical Engineering Reference Manual*.

The *Electrical Engineering Reference Manual* may be purchased from PPI at **www.ppi2pass.com** or from your favorite bookstore.

Table of Contents

The topics and chapters listed here correspond to those in the *Electrical Engineering Reference Manual*, eighth edition.

Preface

The eighth edition of *Electrical Engineering Practice Problems* has been updated to be compatible with the 2008 *National Electrical Code* (NEC). Furthermore, problems covering illumination, C++, and the *National Electrical Safety Code* (NESC) have been added to correspond with the new content in the *Electrical Engineering Reference Manual*. The additions to this book reflect a continued effort to align *Practice Problems* with NCEES exam specification changes.

Electrical Engineering Practice Problems owes much to its predecessors and to the outstanding guidance of PPI.

When you work the problems presented here, you'll be preparing for the NCEES Principles and Practice of Engineering (PE) examination in electrical and computer engineering in the most efficient way possible. This book explores the exam topics systematically, giving you a comprehensive review. The problems are written to suggest areas of further study, the incorrect answers themselves can enlighten, and the correct answers are thoroughly documented. When you take full advantage of the practice problems and solutions, you'll gain knowledge essential for peak performance on your exam module.

If you're a student of electrical engineering, you'll find this book to be an all-inclusive review of the fundamentals of electrical engineering—useful regardless of your current topic of interest. If you're a practicing engineer, you'll find that the logic followed in the solutions can help you as you attempt to resolve similar everyday engineering problems.

Reading the *Electrical Engineering Reference Manual for the Power, Electrical and Electronics, and Computer PE Exams* in conjunction with this text will greatly enhance your study efforts. This *Practice Problems* book is written as a companion to the *Reference Manual*, and the subjects covered in the two books correspond chapter for chapter. The solutions given here often cite the *Reference Manual*, so it would be helpful to have a copy on hand for quick access to background information, including equations, figures, charts, and tables of data. The *Reference Manual* includes many essential appendices of mathematical, basic theoretical, and practical information you can refer to while solving problems. It also features a glossary of commonly used electrical terms and an in-depth index to aid in the search for specific information.

While this book covers all the topics on the PE examination, the focus of the book is basic electrical engineering. With an understanding of these basics, you will better understand the majority of examination questions, as well as real-world electrical engineering (EE) problems. With an in-depth understanding of the basics, problems can be segregated into comprehensible sections that provide an answer to a complex problem when combined.

The sources used in assembling the content of this book include: (1) information on the electrical and computer PE exam made public by the National Council of Examiners for Engineering and Surveying (NCEES); (2) PE exam review course material published by the National Society of Professional Engineers; (3) electrical engineering curricula at leading colleges and universities; (4) current literature in the field of electrical engineering; (5) information on the numerous electrical engineering websites on the Internet; and (6) survey comments from those who have recently taken the PE exam and/or purchased copies of previous editions of this text.

Should you find an error in this book, know that it is mine, and that I regret it. Beyond that, I hope two things happen. First, please let me know about it, by using the errata submission form section on the PPI website at **www.ppi2pass.com/errata**. Second, I hope you learn something from the error—I know I will! I would appreciate constructive comments and suggestions for improvement, additional questions, and recommendations for expansion so that new editions or similar texts will more nearly meet the needs of future examinees.

John A. Camara, PE

Acknowledgments

It is with enduring gratitude that I thank Michael Lindeburg who, with the sixth edition of the *Electrical Engineering Reference Manual*, which this book supplements, gave me the opportunity to realize my dream of authorship of a significant engineering text. It is my hope that this book will match the comprehensive and cohesive quality of the best of Professional Publications' family of engineering books, which I have admired for more than two decades.

The National Society of Professional Engineers provided questions from previous professional engineering examination review courses for my study and guidance. Michael Lindeburg provided Chapters 1–14 on mathematics, as well as the chapters on economic analysis, law, ethics, and PE registration. This saved considerable time. Additionally, there was little, if anything, I could do to improve them, so I'm grateful for their use. Sections of the new Illumination chapter are courtesy of Michael Lindeburg from his outstanding text, the *Engineer-In-Training Reference Manual*. The late Raymond Yarbrough's fifth edition provided the basis for expansion, and numerous figures, tables, and problems were taken, with minor modifications, directly from his book. Steven B. Edington reviewed my initial Communication topic outline and greatly improved its scope and timeliness. John Goularte, Jr., taught me first-hand the applications of biomedical engineering, and the real-world practicality of the questions in this topic is attributable to him. Becky Owens provided guidance on computer fundamentals and showed me practical applications, as well as typed the majority of the index. More importantly, she was patient and supportive through the entire process, even when I wasn't.

Gregg Wagener, PE, reviewed each of the chapters, providing valuable insight, correcting numerous errors and, in general, teaching me a great deal along the way.

A very special thanks to my son, Jac Camara, whose annotated periodic table became the basis for the Electrical Materials topic. Additionally, our discussions about quantum mechanics and a variety of other topics helped stir the intellectual curiosity I thought was fading.

Thanks also to my daughter, Cassiopeia, who has been a constant source of inspiration, challenge, and pride.

Edition 8 thanks go to the very professional and dedicated team at PPI that includes Sarah Hubbard, director of new product development, whose guidance from the beginning has been most appreciated; Cathy Schrott, director of production, who keeps it all moving forward; the style gurus Heather Kinser and Jenny Lindeburg; Tom Bergstrom, technical illustrator; Kate Hayes, the vitally important typesetter; Amy Schwertman, who did the cover design; and Courtnee Crystal, editorial assistant, who put the finishing touches on all.

A special thanks to Sarina Bromberg, Ph.D., whose expertise and challenging questions sharpened the text in numerous chapters. A thanks also to Larry Ota, from The Boeing Company, who reviewed the C++ addition to the text, greatly increasing my knowledge of the subject.

The knowledge expressed in this book represents years of training, instruction, and self-study in electrical, electronic, mechanical, nuclear, marine, space, and a variety of other types of engineering—all of which I find fascinating. I would not understand any of it were it not for the teachers, instructors, mentors, family, and friends who've spent countless hours with me and from whom I've learned so much. A few of them include: Mrs. Mary Avila, Capt. C.E. Ellis, Mr. Claude Estes, Capt. Karl Hasslinger, Mr. Jerome Herbeck, Mr. Jack Hunnicutt, Mr. Ralph Loya, Harry Lynch, Mr. Harold Mackey, Michael O'Neal, Mark Richwine, Mr. Charles Taylor, Mrs. Abigail Thyarks, and Mr. Jim Triguerio. There are others who have touched my life in special ways: Jim, Marla, Lora, Todd, Tom, and Rick. There are many others, though here they will remain nameless, to whom I am indebted. Thanks.

Finally, thanks to my mom, for without her steadfast love little in this world would be worthwhile.

John A. Camara, PE

Codes, Handbooks, and References

This edition of the *Electrical Engineering Practice Problems for the Power, Electrical and Electronics, and Computer Engineering PE Exams* is based on the following codes, handbooks, and references. The most current versions or editions available were used. However, the PE examination is not always based on the most currently available codes, as adoption by state and local controlling authorities often lags issuance by several years.

The PPI website (**www.ppi2pass.com/eefaqs**) provides the versions of the codes, standards, and regulations NCEES has announced the current exams are based. Use this information to decide which editions of these books should be part of your exam preparation.

The minimum recommended library for the electrical exam consists of this book, the National Electrical Code, a standard handbook of electrical engineering, and two textbooks that cover fundamental circuit theory (both electrical and electronic). Numerous textbooks covering basic electrical engineering topics are listed. These texts, or their equivalent (see the topics in brackets), should be used in preparation for the examination. Adequate preparation, not an extensive portable library, is the key to success. Use the latest NCEES requirements and the information in the introduction section of this text as a guide for the necessary references to study for, and take to, the examination.

Codes

Federal Communications Commission (FCC). *Code of Federal Regulations*, "FCC Rules": Title 47-Telecommunications, Ch. 73. "Radio Broadcast Rules," (47CFR73). [Communications]

Institute of Electrical and Electronics Engineers (IEEE). *National Electrical Safety Code*, (NESC). [Power]

National Fire Protection Association (NFPA). *National Electrical Code*, (NFPA 70). [Power]

Standards Organizations of Interest:

ANSI: American National Standards Institute

EIA: Electronic Industries Alliance

FCC: Federal Communications Commission

IEEE: Institute of Electrical and Electronic Engineers

ISA: Instrument Society of America

ISO: International Organization for Standardization[1]

NEMA: National Electrical Manufacturers Association

Handbooks

American Electrician's Handbook. Terrell Croft and Wilford I. Summers. McGraw-Hill. [Power]

CRC Materials Science and Engineering Handbook. James F. Shackelford, William Alexander, and Jun S. Park, eds. CRC Press, Inc. [General]

Electronic Engineer's Handbook. Donald Christiansen, ed. McGraw-Hill. [Electrical and Electronics]

McGraw-Hill Internetworking Handbook. Ed Taylor. McGraw-Hill. [Computers]

National Electrical Code Handbook. Mark W. Earley, Joseph V. Sheehan, et al. National Fire Protection Association. [Power]

National Electrical Safety Code (NESC) *Handbook.* David J. Marie. McGraw-Hill. [Power]

Power System Analysis. John J. Grainger and William D. Stevenson, Jr. McGraw-Hill. [Power]

Process/Industrial Instruments and Controls Handbook. Gregory K. McMillan, ed. McGraw-Hill. [Power and Electrical and Electronics]

Standard Handbook for Electrical Engineers. Donald G. Fink and H. Wayne Beaty. McGraw-Hill. [Power and Electrical and Electronics]

The Biomedical Engineering Handbook. Joseph D. Broizmo, ed. CRC Press. [Electrical and Electronics]

The Communications Handbook. Jerry D. Gibson, ed. CRC Press, Inc. [Electrical and Electronics]

The Computer Science and Engineering Handbook. Allen B. Tucker, Jr., ed. CRC Press, Inc. [Computers]

Wireless and Cellular Telecommunications. William C.Y. Lee. McGraw-Hill. [Electrical and Electronics]

The IESNA Lighting Handbook, Reference & Applications. Mark S. Rea, ed. Illuminating Engineering Society of North America. [Power]

---

[1]ISO is not the acronym for the organization. Rather, "iso" is from the Greek word for "equal."

References

CRC Standard Mathematical Tables. William H. Beyer, ed. CRC Press, Inc. [General]

McGraw-Hill Dictionary of Scientific and Technical Terms. Sybil P. Parker, ed. McGraw-Hill. [General]

Schaum's Outline Series (Electronics and Electrical Engineering). McGraw-Hill. [General]

The Internet for Scientists and Engineers, Brian J. Thomas. SPIE Press & IEEE Press. [Computers]

Texts

An Introduction to Digital and Analog Integrated Circuits and Applications. Sanjit K. Mitra. Harper & Row, Publishers. [Digital Circuit Fundamentals]

Applied Electromagnetics. Martin A. Plonus. McGraw-Hill. [Electromagnetic Theory]

Introduction to Computer Engineering. Taylor L. Booth. John Wiley & Sons. [Computer Design Basics]

Electrical Power Technology. Theodore Wildi. John Wiley & Sons. [Power Theory and Application]

Linear Circuits. M.E. Van Valkenburg and B.K. Kinariwala. Prentice-Hall, Inc. [AC/DC Fundamentals]

Microelectric Circuit Design. Richard C. Jaeger. McGraw-Hill. [Electronic Fundamentals]

Microelectronics. Jacob Millman. McGraw-Hill. [Electronic Fundamentals]

NEC Answers. Michael A. Anthony. McGraw-Hill. [National Electrical Code Example Applications]

How to Use This Book

This book is meant as a companion to the *Electrical Engineering Reference Manual*—a compendium of the electrical engineering field. Admittedly, despite its size, the *Reference Manual* can not cover all the details of electrical engineering. Nevertheless, a thorough study of that volume and its equations, terms, diagrams, and tables, as well as of the problem statements and solutions given in this *Practice Problems* book, will expose you to many electrical engineering topics. This broad overview of the field can benefit you in a number of ways, not the least of which is by thoroughly preparing you for the electrical and computer engineering PE exam. Further, the range of subjects covered allows you to tailor your review toward the topics of your choice, and to study at a level that suits your needs.

Before you start studying, read about the format and content of the PE exam in the *Electrical Engineering Reference Manual* or on the Professional Publications website at **www.ppi2pass.com/eefaqs**. Then, begin your studies with the fundamentals and progressively expand your review to include more specific topics. Keep in mind that *experience in problem solving is the key to success on the exam.*

Depending on your specific study methods, you may wish to study the "easier" material first to build confidence. Alternately, you might delay reading the easy material until just prior to the exam. The choice is yours.

If you are using this text as a review of a particular area of study, go directly to the topic of interest and begin. Any weakness perceived while attempting to solve the problems should prompt you to review the applicable topic in the *Reference Manual*.

Allow yourself three to six months to prepare. You'll get the most benefit out of your exam preparation efforts if you make a plan and stick to it. Reserve time to review problems from topics you've read, up until the examination date, in order to maintain maximum proficiency.

I'm confident that with diligent study, you will find the electrical engineering information you seek within these pages or within the pages of the *Reference Manual*. At the very least, these companion books will allow you to conduct an intelligent search for information. I hope they serve you well. Enjoy the adventure of learning!

John A. Camara, PE

1 Systems of Units

PRACTICE PROBLEMS

1. Convert 250°F to degrees Celsius.
- (A) 115°C
- (B) 121°C
- (C) 124°C
- (D) 420°C

2. Convert the Stefan-Boltzmann constant (0.1713×10^{-8} Btu/ft^2-hr-°R^4) from customary U.S. to SI units.
- (A) 5.14×10^{-10} W/m^2·K^4
- (B) 0.950×10^{-8} W/m^2·K^4
- (C) 5.67×10^{-8} W/m^2·K^4
- (D) 7.33×10^{-6} W/m^2·K^4

3. How many U.S. tons (2000 lbm per ton) of coal with a heating value of 13,000 Btu/lbm must be burned to provide as much energy as a complete nuclear conversion of 1 g of its mass? (Hint: Use Einstein's equation: $E = mc^2$.)
- (A) 1.7 tons
- (B) 14 tons
- (C) 780 tons
- (D) 3300 tons

4. What is the SI unit for force, N, in terms of more basic units?
- (A) kg·m/s^2
- (B) kg·m^2/s^2
- (C) kg·m^2/s^3
- (D) kg/m·s^2

5. What is the appropriate SI unit for **B**?
- (A) henry, H
- (B) siemens, S
- (C) tesla, T
- (D) weber, W

SOLUTIONS

1. The conversion of temperature Fahrenheit (T_F) to Celsius (T_C) is

$$T_C = \left(\tfrac{5}{9}\right)(T_F - 32)$$
$$= \left(\tfrac{5}{9}\right)(250°F - 32°F)$$
$$= \left(\tfrac{5}{9}\right)(218°F)$$
$$= \boxed{121.1°C \quad (121°C)}$$

The answer is (B).

2. In customary U.S. units, the Stefan-Boltzmann constant, σ, is 0.1713×10^{-8} Btu/hr-ft^2-°R^4.

Use the following conversion factors.

$$1 \text{ Btu/hr} = 0.2931 \text{ W}$$
$$1 \text{ ft} = 0.3048 \text{ m}$$
$$T_R = \tfrac{9}{5}T_K$$

Performing the conversion gives

$$\sigma = \left(0.1713 \times 10^{-8}\frac{\text{Btu}}{\text{hr-ft}^2\text{-°R}^4}\right)\left(0.2931\frac{\text{W}}{\frac{\text{Btu}}{\text{hr}}}\right)$$
$$\times \left(\frac{1 \text{ ft}}{0.3048 \text{ m}}\right)^2\left(\frac{9}{5}\frac{°R}{K}\right)^4$$
$$= \boxed{5.67 \times 10^{-8} \text{ W/m}^2\text{·K}^4}$$

The answer is (C).

3. The energy produced from the nuclear conversion of any quantity of mass is given as

$$E = mc^2$$

The speed of light, c, is 3×10^8 m/s.

For a mass of 1 g (0.001 kg),

$$E = mc^2$$

$$= (0.001 \text{ kg}) \left(3 \times 10^8 \ \frac{\text{m}}{\text{s}}\right)^2$$

$$= 9 \times 10^{13} \text{ J}$$

Convert to U.S. customary units with the conversion 1 Btu = 1055 J.

$$E = (9 \times 10^{13} \text{ J}) \left(\frac{1 \text{ Btu}}{1055 \text{ J}}\right)$$

$$= 8.53 \times 10^{10} \text{ Btu}$$

The number, n, of tons of 13,000 Btu/lbm coal is

$$n \text{ tons} = \frac{8.53 \times 10^{10} \text{ Btu}}{\left(13{,}000 \ \frac{\text{Btu}}{\text{lbm}}\right)\left(2000 \ \frac{\text{lbm}}{\text{ton}}\right)}$$

$$= \boxed{3281 \text{ tons} \quad (3300 \text{ tons})}$$

The answer is (D).

4. The newton, N, is the SI unit of force. Using primary dimensional analysis,

$$F = \frac{ML}{\theta^2}$$

$$= \boxed{\frac{\text{kg·m}}{\text{s}^2}}$$

The answer is (A).

5. The SI unit for magnetic flux density, **B**, is the tesla, T.

The answer is (C).

2 Energy, Work, and Power

PRACTICE PROBLEMS

1. A volume of 40×10^6 m³ of water behind a dam is at an average height of 100 m above the input to electrical turbines. The turbines are 90% efficient. What is the approximate maximum potential energy converted to electrical energy?

(A) 3.9×10^{11} J
(B) 3.5×10^{13} J
(C) 3.9×10^{13} J
(D) 1.3×10^{14} J

2. A certain motor parameter is measured by the following expression.

$$\int \mathbf{T} \cdot d\boldsymbol{\theta}$$

What is the motor parameter being measured?

(A) power
(B) thrust
(C) torque
(D) work

3. A linear spring, held by an electric coil, is designed to open a cooling valve in the event of an electrical failure. The conditions are as shown.

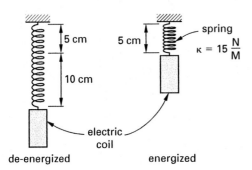

What work must the electric coil do to keep the spring compressed?

(A) 0.018 N·m
(B) 0.075 N·m
(C) 750 N·m
(D) 75×10^3 N·m

4. Consider the illustration shown. An induction motor is rotating at 1800 rpm. The mass is concentrated on the circumference of the rotor. What is the approximate kinetic energy of the rotor?

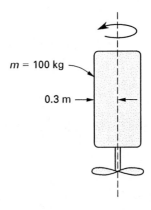

(A) 2.8×10^3 J
(B) 16×10^4 J
(C) 32×10^4 J
(D) 6.4×10^5 J

5. A rotating motor is calculated to have 20×10^4 J of rotational energy during operation. During fault analysis, this motor is expected to dissipate all this energy in one second through a ground. What power is provided to the ground site by the motor?

(A) 20 W
(B) 200 W
(C) 200 kW
(D) 400 kW

SOLUTIONS

1. Using the average height of 100 m, the potential energy, $E_{available}$, is given by the product of the mass of the water (its density, ρ_{H_2O}, times its volume), the acceleration of gravity, g, and the height of the dam above the turbines, z.

$$
\begin{aligned}
E_{available} &= mgz \\
&= (\rho_{H_2O} V_{dam})gz \\
&= \left(1000 \ \frac{kg}{m^3}\right)(40 \times 10^6 \ m^3) \\
&\quad \times \left(9.81 \ \frac{m}{s^2}\right)(100 \ m) \\
&= 3.924 \times 10^{13} \ J
\end{aligned}
$$

At an efficiency, η, of 90%, the maximum energy output is

$$
\begin{aligned}
E_{output} &= \eta E_{available} \\
&= (0.90)(3.924 \times 10^{13} \ J) \\
&= \boxed{3.53 \times 10^{13} \ J \quad (3.5 \times 10^{13} \ J)}
\end{aligned}
$$

The answer is (B).

2. The expression represents the work done in a rotational system by a variable torque.

$$
W = \int \mathbf{T} \cdot d\theta
$$

The answer is (D).

3. Because no value is given for the coil efficiency, assume 100%. Therefore, the work of the coil, W_{coil}, equals the work of a spring, W_{spring}, with spring constant, κ, and positions, $\delta_{initial}$ and δ_{final}.

$$
W_{coil} = W_{spring} = \tfrac{1}{2}\kappa\left(\delta_{final}^2 - \delta_{initial}^2\right)
$$

With no initial deflection, the equation becomes

$$
\begin{aligned}
W_{spring} &= \tfrac{1}{2}\kappa\delta_{final}^2 \\
&= \left(\frac{1}{2}\right)\left(15 \ \frac{N}{m}\right)(10 \ cm)^2 \left(\frac{1 \ m}{100 \ cm}\right)^2 \\
&= \boxed{0.075 \ N \cdot m}
\end{aligned}
$$

The answer is (B).

4. The kinetic energy of the rotor is

$$
E = \tfrac{1}{2}mv^2 \qquad \text{[I]}
$$

The mass is m and the velocity is v.

The velocity is the arclength, s, traveled per unit time.

$$
\begin{aligned}
v &= \frac{s}{t} \\
&= \left(2\pi r \ \frac{m}{rev}\right)\left(\frac{rev}{min}\right) \\
&= 2\pi(0.3 \ m)\left(1800 \ \frac{rev}{min}\right)\left(\frac{1 \ min}{60 \ s}\right) \\
&= 56.55 \ m/s \qquad \text{[II]}
\end{aligned}
$$

Substitute the result of Eq. II into Eq. I, along with the given information about the mass.

$$
\begin{aligned}
E &= \tfrac{1}{2}mv^2 \\
&= \left(\frac{1}{2}\right)(100 \ kg)\left(56.55 \ \frac{m}{s}\right)^2 \\
&= \boxed{15.98 \times 10^4 \ J \quad (16 \times 10^4 \ J)}
\end{aligned}
$$

This problem may also be solved using moment of inertia concepts.

The answer is (B).

5. During worst-case conditions, a motor's rotational energy will be converted to electrical energy and provided to the fault site. Power, in watts, is energy per unit time, or

$$
\begin{aligned}
P &= \frac{E}{t} \\
&= \frac{20 \times 10^4 \ J}{1 \ s} \\
&= \boxed{20 \times 10^4 \ W \quad (200 \ kW)}
\end{aligned}
$$

The answer is (C).

3 Engineering Drawing Practice

PRACTICE PROBLEMS

1. Which of the following symbols represents an electrical connection, or tie point, between one part of an electrical circuit and another?

(A) ⟶▷

(B) ⟶⊣⊢

(C) ⟶⊱

(D) ⟩⟨⟨

2. Which symbol represents a ground for safety purposes on a metallic enclosure?

(A) ⟶▷

(B) ⟶⊣⊢

(C) ⟶⊱

(D) ⟶◡◠◡⟶

3. Which of the following symbols would be used in an electrical schematic to represent a MOSFET?

(A)

(B)

(C)

(D)

4. Which of the following symbols would be used in an electrical schematic to represent a NPN bipolar junction transistor?

(A)

(B)

(C)

(D)

5. Which of the following symbols represents a controlled source?

(A) ⟶⊣⊢⟶

(B) ⟶◯⟶

(C) ⟶◇⟶

(D)

SOLUTIONS

1. A tie-point is represented in an electrical schematic by the symbol

The answer is (A).

2. An earth ground is used for safety. The symbol for an earth ground is

The answer is (B).

3. An *n*-channel enhancement (four-terminal) MOS-FET is symbolized by

The answer is (D).

4. An NPN bipolar junction transistor is represented by

A useful memory aid is that an NPN transistor's arrow is "<u>n</u>ot pointing <u>in</u>" toward the base, thus NPN.

The answer is (A).

5. A controlled source is used in schematics of electronic models, and is represented by the diamond shape.

The answer is (C).

4 Algebra

PRACTICE PROBLEMS

Series

1. Calculate the following sum.

$$\sum_{j=1}^{5}\left((j+1)^2 - 1\right)$$

(A) 15
(B) 24
(C) 35
(D) 85

Logarithms

2. If every 0.1 sec a quantity increases by 0.1% of its current value, calculate the doubling time.

(A) 14 sec
(B) 70 sec
(C) 690 sec
(D) 69,000 sec

3. What are the roots of the following equation?

$$x^2 + 5x + 4 = 0$$

(A) -1, -4
(B) -1, 4
(C) 1, -4
(D) -11, -14

4. What is the base-10 logarithm of 115^2?

(A) 2.06
(B) 4.12
(C) 8.24
(D) 16.5

5. What is the exponential form of $\mathbf{z} = 2 + j2$?

(A) $0.7e^{j2.83}$
(B) $e^{j0.78}$
(C) $0.78e^{j}$
(D) $2.83e^{j0.78}$

SOLUTIONS

1. Let $S_n = (j+1)^2 - 1$.

For $j = 1$,
$$S_1 = (1+1)^2 - 1 = 3$$

For $j = 2$,
$$S_2 = (2+1)^2 - 1 = 8$$

For $j = 3$,
$$S_3 = (3+1)^2 - 1 = 15$$

For $j = 4$,
$$S_4 = (4+1)^2 - 1 = 24$$

For $j = 5$,
$$S_5 = (5+1)^2 - 1 = 35$$

Substituting the above expressions gives

$$\sum_{j=1}^{5}\left((j+1)^2 - 1\right) = \sum_{j=1}^{5} S_j$$
$$= S_1 + S_2 + S_3 + S_4 + S_5$$
$$= 3 + 8 + 15 + 24 + 35$$
$$= \boxed{85}$$

The answer is (D).

2. Let n represent the number of elapsed periods of 0.1 sec, and let y_n represent the quantity present after n periods.

y_0 represents the initial quantity.

$$y_1 = 1.001y_0$$
$$y_2 = 1.001y_1 = (1.001)(1.001y_0) = (1.001)^2 y_0$$

Therefore, by deduction,

$$y_n = (1.001)^n y_0$$

The expression for a doubling of the original quantity is

$$2y_0 = y_n$$

Substitute for y_n.

$$2y_0 = (1.001)^n y_0$$

$$2 = (1.001)^n$$

Take the logarithm of both sides.

$$\log 2 = \log 1.001^n$$

$$= n \log 1.001$$

Solve for n.

$$n = \frac{\log 2}{\log 1.001} = 693.5$$

Because each period, p, is 0.1 sec, the time is

$$t = np$$

$$= n(0.1 \text{ sec})$$

$$= (693.5)(0.1 \text{ sec})$$

$$= \boxed{69.35 \text{ sec} \quad (70 \text{ sec})}$$

The answer is (B).

3. The equation is in the quadratic form,

$$ax^2 + bx + c = 0$$

Quadratic equations can represent a second order response by electrical circuits. The roots are

$$x_1, x_2 = \frac{-b \pm \sqrt{b^2 - 4ac}}{2a}$$

$$= \frac{-5 \pm \sqrt{5^2 - (4)(1)(4)}}{(2)(1)}$$

$$= \frac{-5 \pm 3}{2}$$

$$= \boxed{-1, -4}$$

The answer is (A).

4. The property of logarithms that is useful in this case is

$$\log x^a = a \log x$$

Substitute the given information.

$$\log 115^2 = 2 \log 115$$

$$= (2)(2.06)$$

$$= \boxed{4.12}$$

The answer is (B).

5. The exponential form of a complex number **z** is

$$z = re^{j\theta}$$

The radius in the complex plane is r. The square root of -1 is j, and θ is the angle of the radius in the complex plane. Calculate the unknown values using

$$r = \sqrt{x_{\text{real}}^2 + y_{\text{imaginary}}^2}$$

$$= \sqrt{2^2 + 2^2}$$

$$= \sqrt{8}$$

$$= 2.83$$

$$\theta = \arctan \frac{y}{x} = \arctan \frac{2}{2}$$

$$= \arctan 1$$

$$= 0.78 \text{ rad}$$

Therefore,

$$z = re^{j\theta}$$

$$= \boxed{2.83e^{j0.78}}$$

The answer is (D).

5 Linear Algebra

PRACTICE PROBLEMS

1. What is the determinant of matrix **A**?

$$\mathbf{A} = \begin{bmatrix} 8 & 2 & 0 & 0 \\ 2 & 8 & 2 & 0 \\ 0 & 2 & 8 & 2 \\ 0 & 0 & 2 & 4 \end{bmatrix}$$

(A) 459
(B) 832
(C) 1552
(D) 1776

2. Use Cramer's rule to solve for the values of x, y, and z that simultaneously satisfy the following equations.

$$x + y = -4$$
$$x + z - 1 = 0$$
$$2z - y + 3x = 4$$

(A) $(x, y, z) = (3, 2, 1)$
(B) $(x, y, z) = (-3, -1, 2)$
(C) $(x, y, z) = (3, -1, -3)$
(D) $(x, y, z) = (-1, -3, 2)$

3. What value of x satisfies all three given equations?

$$4x + 6y - 8z = 2$$
$$6x - 2y - 4z = 8$$
$$8x - 14y - 12z = -14$$

(A) 0.10
(B) 0.20
(C) 0.33
(D) 3.0

4. What is the product of matrices **A** and **B**?

$$\mathbf{A} = \begin{bmatrix} 1 & 2 & 5 \\ 3 & 4 & 7 \end{bmatrix}$$

$$\mathbf{B} = \begin{bmatrix} 2 & 9 \\ 6 & 2 \\ 3 & 4 \end{bmatrix}$$

(A) $\begin{bmatrix} 5 & 25 \\ 3 & 47 \end{bmatrix}$
(B) $\begin{bmatrix} 29 & 51 \\ 33 & 63 \end{bmatrix}$
(C) $\begin{bmatrix} 29 & 33 \\ 51 & 63 \end{bmatrix}$
(D) can't be multiplied

5. What is the determinant of matrix **A**?

$$\mathbf{A} = \begin{bmatrix} 4 & 4 \\ 3 & 6 \end{bmatrix}$$

(A) −12
(B) 12
(C) 24
(D) 36

SOLUTIONS

1. Expand by cofactors of the first column, because there are two zeros in that column. (The first row could also have been used.)

$$D = 8\begin{vmatrix} 8 & 2 & 0 \\ 2 & 8 & 2 \\ 0 & 2 & 4 \end{vmatrix} - 2\begin{vmatrix} 2 & 0 & 0 \\ 2 & 8 & 2 \\ 0 & 2 & 4 \end{vmatrix} + 0 - 0$$

by first column:

$$\begin{vmatrix} 8 & 2 & 0 \\ 2 & 8 & 2 \\ 0 & 2 & 4 \end{vmatrix} = (8)((8)(4) - (2)(2)) - (2)((2)(4) - (2)(0))$$
$$= 208$$

by first column:

$$\begin{vmatrix} 2 & 0 & 0 \\ 2 & 8 & 2 \\ 0 & 2 & 4 \end{vmatrix} = (2)((8)(4) - (2)(2))$$
$$= 56$$

$$D = (8)(208) - (2)(56) = \boxed{1552}$$

The answer is (C).

2. Rearrange the equations.

$$\begin{array}{rcl} x + y & = & -4 \\ x \quad + z & = & 1 \\ 3x - y + 2z & = & 4 \end{array}$$

Write the set of equations in matrix form: $\mathbf{AX} = \mathbf{B}$.

$$\begin{bmatrix} 1 & 1 & 0 \\ 1 & 0 & 1 \\ 3 & -1 & 2 \end{bmatrix}\begin{bmatrix} x \\ y \\ z \end{bmatrix} = \begin{bmatrix} -4 \\ 1 \\ 4 \end{bmatrix}$$

Find the determinant of the matrix $\mathbf{A}$.

$$|\mathbf{A}| = \begin{vmatrix} 1 & 1 & 0 \\ 1 & 0 & 1 \\ 3 & -1 & 2 \end{vmatrix}$$
$$= 1\begin{vmatrix} 0 & 1 \\ -1 & 2 \end{vmatrix} - 1\begin{vmatrix} 1 & 0 \\ -1 & 2 \end{vmatrix} + 3\begin{vmatrix} 1 & 0 \\ 0 & 1 \end{vmatrix}$$
$$= (1)((0)(2) - (1)(-1))$$
$$\quad - (1)((1)(2) - (-1)(0))$$
$$\quad + (3)((1)(1) - (0)(0))$$
$$= (1)(1) - (1)(2) + (3)(1)$$
$$= 1 - 2 + 3$$
$$= 2$$

Find the determinant of the substitutional matrix $\mathbf{A_1}$.

$$|\mathbf{A_1}| = \begin{vmatrix} -4 & 1 & 0 \\ 1 & 0 & 1 \\ 4 & -1 & 2 \end{vmatrix}$$
$$= -4\begin{vmatrix} 0 & 1 \\ -1 & 2 \end{vmatrix} - 1\begin{vmatrix} 1 & 0 \\ -1 & 2 \end{vmatrix} + 4\begin{vmatrix} 1 & 0 \\ 0 & 1 \end{vmatrix}$$
$$= (-4)((0)(2) - (1)(-1))$$
$$\quad - (1)((1)(2) - (-1)(0))$$
$$\quad + (4)((1)(1) - (0)(0))$$
$$= (-4)(1) - (1)(2) + (4)(1)$$
$$= -4 - 2 + 4$$
$$= -2$$

Find the determinant of the substitutional matrix $\mathbf{A_2}$.

$$|\mathbf{A_2}| = \begin{vmatrix} 1 & -4 & 0 \\ 1 & 1 & 1 \\ 3 & 4 & 2 \end{vmatrix}$$
$$= 1\begin{vmatrix} 1 & 1 \\ 4 & 2 \end{vmatrix} - 1\begin{vmatrix} -4 & 0 \\ 4 & 2 \end{vmatrix} + 3\begin{vmatrix} -4 & 0 \\ 1 & 1 \end{vmatrix}$$
$$= (1)((1)(2) - (4)(1))$$
$$\quad - (1)((-4)(2) - (4)(0))$$
$$\quad + (3)((-4)(1) - (1)(0))$$
$$= (1)(-2) - (1)(-8) + (3)(-4)$$
$$= -2 + 8 - 12$$
$$= -6$$

Find the determinant of the substitutional matrix $\mathbf{A_3}$.

$$|\mathbf{A_3}| = \begin{vmatrix} 1 & 1 & -4 \\ 1 & 0 & 1 \\ 3 & -1 & 4 \end{vmatrix}$$
$$= 1\begin{vmatrix} 0 & 1 \\ -1 & 4 \end{vmatrix} - 1\begin{vmatrix} 1 & -4 \\ -1 & 4 \end{vmatrix} + 3\begin{vmatrix} 1 & -4 \\ 0 & 0 \end{vmatrix}$$
$$= (1)((0)(4) - (-1)(1))$$
$$\quad - (1)((1)(4) - (-1)(-4))$$
$$\quad + (3)((1)(1) - (0)(-4))$$
$$= (1)(1) - (1)(0) + (3)(1)$$
$$= 1 - 0 + 3$$
$$= 4$$

Use Cramer's rule.

$$x = \frac{|\mathbf{A_1}|}{|\mathbf{A}|} = \frac{-2}{2} = \boxed{-1}$$

$$y = \frac{|\mathbf{A_2}|}{|\mathbf{A}|} = \frac{-6}{2} = \boxed{-3}$$

$$z = \frac{|\mathbf{A_3}|}{|\mathbf{A}|} = \frac{4}{2} = \boxed{2}$$

The answer is (D).

5 Linear Algebra

PRACTICE PROBLEMS

1. What is the determinant of matrix $\mathbf{A}$?

$$\mathbf{A} = \begin{bmatrix} 8 & 2 & 0 & 0 \\ 2 & 8 & 2 & 0 \\ 0 & 2 & 8 & 2 \\ 0 & 0 & 2 & 4 \end{bmatrix}$$

(A) 459
(B) 832
(C) 1552
(D) 1776

2. Use Cramer's rule to solve for the values of x, y, and z that simultaneously satisfy the following equations.

$$x + y = -4$$
$$x + z - 1 = 0$$
$$2z - y + 3x = 4$$

(A) $(x, y, z) = (3, 2, 1)$
(B) $(x, y, z) = (-3, -1, 2)$
(C) $(x, y, z) = (3, -1, -3)$
(D) $(x, y, z) = (-1, -3, 2)$

3. What value of x satisfies all three given equations?

$$4x + 6y - 8z = 2$$
$$6x - 2y - 4z = 8$$
$$8x - 14y - 12z = -14$$

(A) 0.10
(B) 0.20
(C) 0.33
(D) 3.0

4. What is the product of matrices $\mathbf{A}$ and $\mathbf{B}$?

$$\mathbf{A} = \begin{bmatrix} 1 & 2 & 5 \\ 3 & 4 & 7 \end{bmatrix}$$

$$\mathbf{B} = \begin{bmatrix} 2 & 9 \\ 6 & 2 \\ 3 & 4 \end{bmatrix}$$

(A) $\begin{bmatrix} 5 & 25 \\ 3 & 47 \end{bmatrix}$

(B) $\begin{bmatrix} 29 & 51 \\ 33 & 63 \end{bmatrix}$

(C) $\begin{bmatrix} 29 & 33 \\ 51 & 63 \end{bmatrix}$

(D) can't be multiplied

5. What is the determinant of matrix $\mathbf{A}$?

$$\mathbf{A} = \begin{bmatrix} 4 & 4 \\ 3 & 6 \end{bmatrix}$$

(A) -12
(B) 12
(C) 24
(D) 36

SOLUTIONS

1. Expand by cofactors of the first column, because there are two zeros in that column. (The first row could also have been used.)

$$D = 8\begin{vmatrix} 8 & 2 & 0 \\ 2 & 8 & 2 \\ 0 & 2 & 4 \end{vmatrix} - 2\begin{vmatrix} 2 & 0 & 0 \\ 2 & 8 & 2 \\ 0 & 2 & 4 \end{vmatrix} + 0 - 0$$

by first column:

$$\begin{vmatrix} 8 & 2 & 0 \\ 2 & 8 & 2 \\ 0 & 2 & 4 \end{vmatrix} = (8)\big((8)(4) - (2)(2)\big) - (2)\big((2)(4) - (2)(0)\big)$$
$$= 208$$

by first column:

$$\begin{vmatrix} 2 & 0 & 0 \\ 2 & 8 & 2 \\ 0 & 2 & 4 \end{vmatrix} = (2)\big((8)(4) - (2)(2)\big)$$
$$= 56$$

$$D = (8)(208) - (2)(56) = \boxed{1552}$$

The answer is (C).

2. Rearrange the equations.

$$\begin{aligned} x + y \quad\ &= -4 \\ x \quad\ + z &= 1 \\ 3x - y + 2z &= 4 \end{aligned}$$

Write the set of equations in matrix form: $\mathbf{AX} = \mathbf{B}$.

$$\begin{bmatrix} 1 & 1 & 0 \\ 1 & 0 & 1 \\ 3 & -1 & 2 \end{bmatrix}\begin{bmatrix} x \\ y \\ z \end{bmatrix} = \begin{bmatrix} -4 \\ 1 \\ 4 \end{bmatrix}$$

Find the determinant of the matrix $\mathbf{A}$.

$$|\mathbf{A}| = \begin{vmatrix} 1 & 1 & 0 \\ 1 & 0 & 1 \\ 3 & -1 & 2 \end{vmatrix}$$
$$= 1\begin{vmatrix} 0 & 1 \\ -1 & 2 \end{vmatrix} - 1\begin{vmatrix} 1 & 0 \\ -1 & 2 \end{vmatrix} + 3\begin{vmatrix} 1 & 0 \\ 0 & 1 \end{vmatrix}$$
$$= (1)\big((0)(2) - (1)(-1)\big)$$
$$\quad - (1)\big((1)(2) - (-1)(0)\big)$$
$$\quad + (3)\big((1)(1) - (0)(0)\big)$$
$$= (1)(1) - (1)(2) + (3)(1)$$
$$= 1 - 2 + 3$$
$$= 2$$

Find the determinant of the substitutional matrix $\mathbf{A_1}$.

$$|\mathbf{A_1}| = \begin{vmatrix} -4 & 1 & 0 \\ 1 & 0 & 1 \\ 4 & -1 & 2 \end{vmatrix}$$
$$= -4\begin{vmatrix} 0 & 1 \\ -1 & 2 \end{vmatrix} - 1\begin{vmatrix} 1 & 0 \\ -1 & 2 \end{vmatrix} + 4\begin{vmatrix} 1 & 0 \\ 0 & 1 \end{vmatrix}$$
$$= (-4)\big((0)(2) - (1)(-1)\big)$$
$$\quad - (1)\big((1)(2) - (-1)(0)\big)$$
$$\quad + (4)\big((1)(1) - (0)(0)\big)$$
$$= (-4)(1) - (1)(2) + (4)(1)$$
$$= -4 - 2 + 4$$
$$= -2$$

Find the determinant of the substitutional matrix $\mathbf{A_2}$.

$$|\mathbf{A_2}| = \begin{vmatrix} 1 & -4 & 0 \\ 1 & 1 & 1 \\ 3 & 4 & 2 \end{vmatrix}$$
$$= 1\begin{vmatrix} 1 & 1 \\ 4 & 2 \end{vmatrix} - 1\begin{vmatrix} -4 & 0 \\ 4 & 2 \end{vmatrix} + 3\begin{vmatrix} -4 & 0 \\ 1 & 1 \end{vmatrix}$$
$$= (1)\big((1)(2) - (4)(1)\big)$$
$$\quad - (1)\big((-4)(2) - (4)(0)\big)$$
$$\quad + (3)\big((-4)(1) - (1)(0)\big)$$
$$= (1)(-2) - (1)(-8) + (3)(-4)$$
$$= -2 + 8 - 12$$
$$= -6$$

Find the determinant of the substitutional matrix $\mathbf{A_3}$.

$$|\mathbf{A_3}| = \begin{vmatrix} 1 & 1 & -4 \\ 1 & 0 & 1 \\ 3 & -1 & 4 \end{vmatrix}$$
$$= 1\begin{vmatrix} 0 & 1 \\ -1 & 4 \end{vmatrix} - 1\begin{vmatrix} 1 & -4 \\ -1 & 4 \end{vmatrix} + 3\begin{vmatrix} 1 & -4 \\ 0 & 0 \end{vmatrix}$$
$$= (1)\big((0)(4) - (-1)(1)\big)$$
$$\quad - (1)\big((1)(4) - (-1)(-4)\big)$$
$$\quad + (3)\big((1)(1) - (0)(-4)\big)$$
$$= (1)(1) - (1)(0) + (3)(1)$$
$$= 1 - 0 + 3$$
$$= 4$$

Use Cramer's rule.

$$x = \frac{|\mathbf{A_1}|}{|\mathbf{A}|} = \frac{-2}{2} = \boxed{-1}$$
$$y = \frac{|\mathbf{A_2}|}{|\mathbf{A}|} = \frac{-6}{2} = \boxed{-3}$$
$$z = \frac{|\mathbf{A_3}|}{|\mathbf{A}|} = \frac{4}{2} = \boxed{2}$$

The answer is (D).

3. Use Cramer's rule to solve for x. The determinant of the coefficient matrix is

$$|\mathbf{A}| = \begin{vmatrix} 4 & 6 & -8 \\ 6 & -2 & -4 \\ 8 & -14 & -12 \end{vmatrix}$$

$$= 4\begin{vmatrix} -2 & -4 \\ -14 & -12 \end{vmatrix} - 6\begin{vmatrix} 6 & -8 \\ -14 & -12 \end{vmatrix} + 8\begin{vmatrix} 6 & -8 \\ -2 & -4 \end{vmatrix}$$

$$= (4)\big((-2)(-12) - (-4)(-14)\big)$$
$$\quad - (6)\big((6)(-12) - (-8)(-14)\big)$$
$$\quad + (8)\big((6)(-4) - (-2)(-8)\big)$$

$$= (4)(24 - 56) - (6)(-72 - 112)$$
$$\quad + (8)(-24 - 16)$$

$$= 656$$

The determinant of the substitutional matrix is

$$|\mathbf{A}_1| = \begin{vmatrix} 2 & 6 & -8 \\ 8 & -2 & -4 \\ -14 & -14 & -12 \end{vmatrix}$$

$$= 1968$$

Therefore,

$$x = \frac{|\mathbf{A}|}{|\mathbf{A}_1|} = \frac{656}{1968} = \frac{1}{3}$$

$$= \boxed{0.33}$$

The answer is (C).

4. Let the product matrix be $\mathbf{C}$.

$$\mathbf{C} = \begin{bmatrix} c_{11} & c_{12} \\ c_{21} & c_{22} \end{bmatrix} = \mathbf{AB}$$

$$= \begin{bmatrix} 1 & 2 & 5 \\ 3 & 4 & 7 \end{bmatrix} \begin{bmatrix} 2 & 9 \\ 6 & 2 \\ 3 & 4 \end{bmatrix}$$

$$c_{11} = (1)(2) + (2)(6) + (5)(3)$$
$$= 2 + 12 + 15$$
$$= 29$$

$$c_{12} = (1)(9) + (2)(2) + (5)(4)$$
$$= 9 + 4 + 20$$
$$= 33$$

$$c_{21} = (3)(2) + (4)(6) + (7)(3)$$
$$= 6 + 24 + 21$$
$$= 51$$

$$c_{22} = (3)(9) + (4)(2) + (7)(4)$$
$$= 27 + 8 + 28$$
$$= 63$$

Thus,

$$\mathbf{C} = \boxed{\begin{bmatrix} 29 & 33 \\ 51 & 63 \end{bmatrix}}$$

The answer is (C).

5. The determinant of any 2×2 matrix is

$$|\mathbf{A}| = \begin{vmatrix} a & c \\ b & d \end{vmatrix} = ad - bc$$

In this case,

$$|\mathbf{A}| = \begin{vmatrix} 4 & 4 \\ 3 & 6 \end{vmatrix} = (4)(6) - (4)(3)$$

$$= 24 - 12$$

$$= \boxed{12}$$

The answer is (B).

6 Vectors

PRACTICE PROBLEMS

1. What is the value of the unit vector for $\mathbf{V} = 2\mathbf{i} + 4\mathbf{j} + 4\mathbf{k}$?
- (A) $1\mathbf{i} + 2\mathbf{j} + 2\mathbf{k}$
- (B) $(1/18)\mathbf{i} + (1/6)\mathbf{j} + (1/6)\mathbf{k}$
- (C) $(2/3)\mathbf{i} + (1/3)\mathbf{j} + (1/3)\mathbf{k}$
- (D) $(1/3)\mathbf{i} + (2/3)\mathbf{j} + (2/3)\mathbf{k}$

2. What is the dot product of the two vectors given?

$$\mathbf{V}_1 = 2\mathbf{i} + 2\mathbf{j} + 4\mathbf{k}$$

$$\mathbf{V}_2 = 2\mathbf{i} + 2\mathbf{j} + 1\mathbf{k}$$

- (A) $4\angle 35°$
- (B) 4
- (C) $12\angle 35°$
- (D) 12

3. Which of the following vectors is orthogonal to $\mathbf{V}_1$ and $\mathbf{V}_2$?

$$\mathbf{V}_1 = \mathbf{i} + \mathbf{j} + 2\mathbf{k}$$

$$\mathbf{V}_2 = \mathbf{i} + 2\mathbf{j}$$

- (A) $-\mathbf{i} - 2\mathbf{j} + 2\mathbf{k}$
- (B) $-4\mathbf{i} + 2\mathbf{j} + 2\mathbf{k}$
- (C) $-4\mathbf{i} - 2\mathbf{j} - \mathbf{k}$
- (D) $4\mathbf{i} - 2\mathbf{j} + \mathbf{k}$

4. The magnitude of the vector cross product corresponds to which of the following?
- (A) area
- (B) line
- (C) projection
- (D) volume

5. Which of the following vectors is orthogonal to $\mathbf{V}_1 = 2\mathbf{i} + 4\mathbf{j}$?
- (A) $2\mathbf{i} - 4\mathbf{j}$
- (B) $4\mathbf{i} - 2\mathbf{j}$
- (C) $-2\mathbf{i} - 4\mathbf{j}$
- (D) $2\mathbf{i} + 4\mathbf{j}$

SOLUTIONS

1. The unit vector $\mathbf{a}$ for vector $\mathbf{V}$ is

$$\mathbf{a} = \frac{\mathbf{V}}{|\mathbf{V}|}$$

$$= \frac{2\mathbf{i} + 4\mathbf{j} + 4\mathbf{k}}{\sqrt{2^2 + 4^2 + 4^2}}$$

$$= \frac{2\mathbf{i} + 4\mathbf{j} + 4\mathbf{k}}{6}$$

$$= \boxed{(1/3)\mathbf{i} + (2/3)\mathbf{j} + (2/3)\mathbf{k}}$$

The answer is (D).

2. The dot product is

$$\mathbf{V}_1 \cdot \mathbf{V}_2 = |\mathbf{V}_1|\,|\mathbf{V}_2|\cos\phi = V_{1x}V_{2x} + V_{1y}V_{2y} + V_{1z}V_{2z}$$

$$= (2)(2) + (2)(2) + (4)(1)$$

$$= 4 + 4 + 4$$

$$= \boxed{12}$$

The answer is (D).

3. An orthogonal vector is found from the cross product. The order of the cross product will change the perpendicular found. Using $\mathbf{V}_1 \times \mathbf{V}_2$ gives

$$\mathbf{V}_1 \times \mathbf{V}_2 = \begin{vmatrix} \mathbf{i} & 1 & 1 \\ \mathbf{j} & 1 & 2 \\ \mathbf{k} & 2 & 0 \end{vmatrix}$$

$$= \mathbf{i}\begin{vmatrix} 1 & 2 \\ 2 & 0 \end{vmatrix} - \mathbf{j}\begin{vmatrix} 1 & 1 \\ 2 & 0 \end{vmatrix} + \mathbf{k}\begin{vmatrix} 1 & 1 \\ 1 & 2 \end{vmatrix}$$

$$= \mathbf{i}(0 - 4) - \mathbf{j}(0 - 2) + \mathbf{k}(2 - 1)$$

$$= \boxed{-4\mathbf{i} + 2\mathbf{j} + 2\mathbf{k}}$$

The answer is (B).

4. The magnitude of the vector cross product for two vectors, say $V_1 \times V_2$, corresponds to the area of a parallelogram with V_1 and V_2 as its sides.

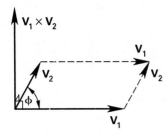

The answer is (A).

5. A vector is orthogonal to another if the dot product of the two vectors equals zero.

Checking answer (A) gives

$$\mathbf{V}_1 \cdot \mathbf{A} = \mathbf{V}_{1x}\mathbf{A}_{1x} + \mathbf{V}_{2y}\mathbf{A}_{2y}$$

$$= (2)(2) + (4)(-4)$$

$$= -12$$

Because the dot product result is not zero, $\mathbf{A}$ is not orthogonal.

Checking answer (B) gives

$$\mathbf{V}_1 \cdot \mathbf{B} = \mathbf{V}_{1x}\mathbf{B}_{1x} + \mathbf{V}_{2y}\mathbf{B}_{2y}$$

$$= (2)(4) + (4)(-2)$$

$$= 8 - 8$$

$$= 0$$

> Thus, $4\mathbf{i} - 2j$ is orthogonal to $\mathbf{V}_1 = 2\mathbf{i} + 4\mathbf{j}$.

The answer is (B).

6 Vectors

PRACTICE PROBLEMS

1. What is the value of the unit vector for $\mathbf{V} = 2\mathbf{i} + 4\mathbf{j} + 4\mathbf{k}$?
- (A) $1\mathbf{i} + 2\mathbf{j} + 2\mathbf{k}$
- (B) $(1/18)\mathbf{i} + (1/6)\mathbf{j} + (1/6)\mathbf{k}$
- (C) $(2/3)\mathbf{i} + (1/3)\mathbf{j} + (1/3)\mathbf{k}$
- (D) $(1/3)\mathbf{i} + (2/3)\mathbf{j} + (2/3)\mathbf{k}$

2. What is the dot product of the two vectors given?

$$\mathbf{V}_1 = 2\mathbf{i} + 2\mathbf{j} + 4\mathbf{k}$$

$$\mathbf{V}_2 = 2\mathbf{i} + 2\mathbf{j} + 1\mathbf{k}$$

- (A) $4\angle 35°$
- (B) 4
- (C) $12\angle 35°$
- (D) 12

3. Which of the following vectors is orthogonal to $\mathbf{V}_1$ and $\mathbf{V}_2$?

$$\mathbf{V}_1 = \mathbf{i} + \mathbf{j} + 2\mathbf{k}$$

$$\mathbf{V}_2 = \mathbf{i} + 2\mathbf{j}$$

- (A) $-\mathbf{i} - 2\mathbf{j} + 2\mathbf{k}$
- (B) $-4\mathbf{i} + 2\mathbf{j} + 2\mathbf{k}$
- (C) $-4\mathbf{i} - 2\mathbf{j} - \mathbf{k}$
- (D) $4\mathbf{i} - 2\mathbf{j} + \mathbf{k}$

4. The magnitude of the vector cross product corresponds to which of the following?
- (A) area
- (B) line
- (C) projection
- (D) volume

5. Which of the following vectors is orthogonal to $\mathbf{V}_1 = 2\mathbf{i} + 4\mathbf{j}$?
- (A) $2\mathbf{i} - 4\mathbf{j}$
- (B) $4\mathbf{i} - 2\mathbf{j}$
- (C) $-2\mathbf{i} - 4\mathbf{j}$
- (D) $2\mathbf{i} + 4\mathbf{j}$

SOLUTIONS

1. The unit vector $\mathbf{a}$ for vector $\mathbf{V}$ is

$$\mathbf{a} = \frac{\mathbf{V}}{|\mathbf{V}|}$$

$$= \frac{2\mathbf{i} + 4\mathbf{j} + 4\mathbf{k}}{\sqrt{2^2 + 4^2 + 4^2}}$$

$$= \frac{2\mathbf{i} + 4\mathbf{j} + 4\mathbf{k}}{6}$$

$$= \boxed{(1/3)\mathbf{i} + (2/3)\mathbf{j} + (2/3)\mathbf{k}}$$

The answer is (D).

2. The dot product is

$$\mathbf{V}_1 \cdot \mathbf{V}_2 = |\mathbf{V}_1||\mathbf{V}_2|\cos\phi = V_{1x}V_{2x} + V_{1y}V_{2y} + V_{1z}V_{2z}$$

$$= (2)(2) + (2)(2) + (4)(1)$$

$$= 4 + 4 + 4$$

$$= \boxed{12}$$

The answer is (D).

3. An orthogonal vector is found from the cross product. The order of the cross product will change the perpendicular found. Using $\mathbf{V}_1 \times \mathbf{V}_2$ gives

$$\mathbf{V}_1 \times \mathbf{V}_2 = \begin{vmatrix} \mathbf{i} & 1 & 1 \\ \mathbf{j} & 1 & 2 \\ \mathbf{k} & 2 & 0 \end{vmatrix}$$

$$= \mathbf{i}\begin{vmatrix} 1 & 2 \\ 2 & 0 \end{vmatrix} - \mathbf{j}\begin{vmatrix} 1 & 1 \\ 2 & 0 \end{vmatrix} + \mathbf{k}\begin{vmatrix} 1 & 1 \\ 1 & 2 \end{vmatrix}$$

$$= \mathbf{i}(0 - 4) - \mathbf{j}(0 - 2) + \mathbf{k}(2 - 1)$$

$$= \boxed{-4\mathbf{i} + 2\mathbf{j} + 2\mathbf{k}}$$

The answer is (B).

4. The magnitude of the vector cross product for two vectors, say $\mathbf{V}_1 \times \mathbf{V}_2$, corresponds to the area of a parallelogram with $\mathbf{V}_1$ and $\mathbf{V}_2$ as its sides.

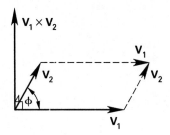

The answer is (A).

5. A vector is orthogonal to another if the dot product of the two vectors equals zero.

Checking answer (A) gives

$$\mathbf{V}_1 \cdot \mathbf{A} = \mathbf{V}_{1x}\mathbf{A}_{1x} + \mathbf{V}_{2y}\mathbf{A}_{2y}$$
$$= (2)(2) + (4)(-4)$$
$$= -12$$

Because the dot product result is not zero, $\mathbf{A}$ is not orthogonal.

Checking answer (B) gives

$$\mathbf{V}_1 \cdot \mathbf{B} = \mathbf{V}_{1x}\mathbf{B}_{1x} + \mathbf{V}_{2y}\mathbf{B}_{2y}$$
$$= (2)(4) + (4)(-2)$$
$$= 8 - 8$$
$$= 0$$

Thus, $4\mathbf{i} - 2j$ is orthogonal to $\mathbf{V}_1 = 2\mathbf{i} + 4\mathbf{j}$.

The answer is (B).

7 Trigonometry

PRACTICE PROBLEMS

1. A 5 lbm (5 kg) block sits on a 20° incline without slipping.

(a) Draw the free-body diagram with respect to axes parallel and perpendicular to the surface of the incline.

(b) Determine the magnitude of the frictional force on the block.

    (A) 1.71 lbf (16.8 N)
    (B) 3.35 lbf (32.9 N)
    (C) 4.70 lbf (46.1 N)
    (D) 5.00 lbf (49.1 N)

2. Consider the unit circle shown.

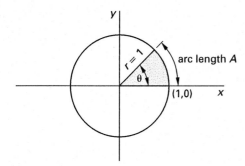

What is the value of the arc length, A, in radians, if the shaded area is $\theta/2$?

    (A) θ
    (B) $\pi/4$
    (C) $\pi/2$
    (D) $r\theta$

3. If the cosine of the angle θ is given as C, what is the tangent of the angle θ?

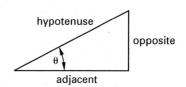

    (A) $\dfrac{C \times \text{hypotenuse}}{\text{opposite}}$

    (B) $\dfrac{\text{adjacent}}{\text{opposite}}$

    (C) $\dfrac{\text{opposite} \times \text{adjacent}}{C \times \text{hypotenuse}}$

    (D) $\dfrac{\text{opposite}}{C \times \text{hypotenuse}}$

4. What is the value of the sum of the squares of the sine and cosine of an angle θ?

    (A) 0
    (B) $\pi/4$
    (C) 1
    (D) 2π

5. What is the approximate value of $\tan 10°$ in radians?

    (A) 1.00
    (B) 10.0°
    (C) 0.0500
    (D) 0.175

SOLUTIONS

1. (a)

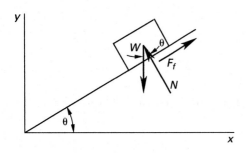

Customary U.S. Solution

(b) The mass of the block, m, is 5 lbm.

The angle of inclination, θ, is 20°. The weight is

$$W = \frac{mg}{g_c}$$

$$= \frac{(5 \text{ lbm})\left(32.2 \dfrac{\text{ft}}{\text{sec}^2}\right)}{32.2 \dfrac{\text{lbm-ft}}{\text{lbf-sec}^2}}$$

$$= 5 \text{ lbf}$$

The frictional force is

$$F_f = W \sin\theta$$
$$= (5 \text{ lbf}) \sin 20°$$
$$= \boxed{1.71 \text{ lbf}}$$

The answer is (A).

SI Solution

(b) The mass of the block, m, is 5 kg.

The angle of inclination, θ, is 20°. The gravitational force is

$$W = mg$$
$$= (5 \text{ kg})\left(9.81 \dfrac{\text{m}}{\text{s}^2}\right)$$
$$= 49.1 \text{ N}$$

The frictional force is

$$F_f = W \sin\theta$$
$$= (49.1 \text{ N}) \sin 20°$$
$$= \boxed{16.8 \text{ N}}$$

The answer is (A).

2. The area of a sector with a central angle of θ radians is $\theta/2$ units for a unit circle. The arc length is therefore θ radians.

The answer is (A).

3. The cosine is

$$C = \frac{\text{adjacent}}{\text{hypotenuse}} \qquad \text{[I]}$$

The tangent is

$$T = \frac{\text{opposite}}{\text{adjacent}} \qquad \text{[II]}$$

Substituting the value of adjacent from Eq. I into Eq. II gives

$$T = \frac{\text{opposite}}{\text{adjacent}} = \boxed{\frac{\text{opposite}}{C \times \text{hypotenuse}}}$$

The answer is (D).

4. For any angle θ, the following is true.

$$\cos^2 + \sin^2 = \boxed{1}$$

The answer is (C).

5. For angles less than approximately 10°, the following is valid.

$$\sin\theta \approx \tan\theta \approx \theta\big|_{\theta \le 10°}$$

The value of θ must be expressed in radians. Converting the 10° to radians gives

$$\theta_{\text{radians}} = 10°\left(\frac{2\pi \text{ rad}}{360°}\right)$$

$$= 0.05\pi \text{ rad}$$

$$= \boxed{0.175 \text{ rad}}$$

The answer is (D).

8 Analytic Geometry

PRACTICE PROBLEMS

1. The diameter of a sphere and of the base of a cone are equal. What percentage of that diameter must the cone's height be, so that both volumes are equal?

(A) 133%
(B) 150%
(C) 166%
(D) 200%

2. Consider the graph shown. What is the slope of the line?

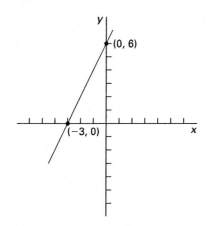

(A) −1/2
(B) 1/2
(C) 2
(D) 4

3. What type of symmetry is shown?

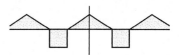

(A) even
(B) odd
(C) rotational
(D) quadratic

4. What expression in the spherical coordinate system represents the z coordinate in the Cartesian coordinate system?

(A) z
(B) $r \sin \theta$
(C) $r \cos \phi$
(D) $r \sin \phi \sin \theta$

5. What geometric figure does the following equation represent?

$$(x - 2)^2 + (y - 2)^2 + (z - 2)^2 = 100$$

(A) circle
(B) ellipse
(C) helix
(D) sphere

SOLUTIONS

1. Let d be the diameter of the sphere and the base of the cone.

The volume of the sphere is

$$V_{\text{sphere}} = \frac{4}{3}\pi r^3 = \frac{4}{3}\pi\left(\frac{d}{2}\right)^3$$
$$= \frac{\pi}{6}d^3$$

The volume of the circular cone is

$$V_{\text{cone}} = \frac{1}{3}\pi r^2 h = \frac{1}{3}\pi\left(\frac{d}{2}\right)^2 h$$
$$= \frac{\pi}{12}d^2 h$$

The volumes of the sphere and cone are equal.

$$V_{\text{cone}} = V_{\text{sphere}}$$
$$\frac{\pi}{12}d^2 h = \frac{\pi}{6}d^3$$
$$h = 2d$$

The cone's height must be 200% of its diameter.

The answer is (D).

2. The two-point form of a line is

$$\frac{y - y_1}{x - x_1} = \frac{y_2 - y_1}{x_2 - x_1}$$

Substitute for the two points given.

$$\frac{y - 6}{x - 0} = \frac{0 - 6}{-3 - 0}$$
$$\frac{y - 6}{x} = 2$$
$$y - 6 = 2x$$
$$y = 2x + 6$$

The equation is in the slope intercept form of a line,

$$y = mx + b$$

The slope value is m. Therefore, in this case, the slope is 2.

The answer is (C).

3. The symmetry is such that the value of y is identical for any $\pm x$ value. This is even symmetry.

The answer is (A).

4. In the spherical system

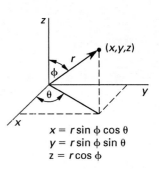

$$x = r\sin\phi\cos\theta$$
$$y = r\sin\phi\sin\theta$$
$$z = r\cos\phi$$

The value of z is $r\cos\phi$.

The answer is (C).

5. The general equation for a sphere centered at (h, k, l) is

$$(x - h)^2 + (y - k)^2 + (z - l)^2 = r^2$$

The equation given is in this form. The equation thus represents a sphere centered at $(2,2,2)$.

The answer is (D).

9 Differential Calculus

PRACTICE PROBLEMS

1. What are the values of a, b, and c in the following expression, such that $n(\infty) = 100$, $n(0) = 10$, and $dn(0)/dt = 0.5$?

$$n(t) = \frac{a}{1 + be^{ct}}$$

(A) $a = 10$, $b = 9$, $c = 1$
(B) $a = 100$, $b = 10$, $c = 1.5$
(C) $a = 100$, $b = 9$, $c = -0.056$
(D) $a = 1000$, $b = 10$, $c = -0.056$

2. Find all minima, maxima, and inflection points for

$$y = x^3 - 9x^2 - 3$$

(A)   maximum at $x = 0$
      inflection at $x = 3$
      minimum at $x = 6$
(B)   maximum at $x = 0$
      inflection at $x = -3$
      minimum at $x = -6$
(C)   minimum at $x = 0$
      inflection at $x = 3$
      maximum at $x = 6$
(D)   minimum at $x = 3$
      inflection at $x = 0$
      maximum at $x = -3$

3. What is the derivative of the following equation?

$$f(t) = 170 \cos 377t$$

(A) $120 \sin 377t$
(B) $170 \sin 377t$
(C) $-170 \sin 377t$
(D) $-64{,}000 \sin 377t$

4. What is the gradient of the following function?

$$f(x, y) = x^2 - 2y^2 + 2x + y$$

(A) $(2x - 2)\mathbf{i} - (4y + 1)\mathbf{j}$
(B) $(2x + 2)\mathbf{i} - (-4y + 1)\mathbf{j}$
(C) $(-2x - 2)\mathbf{i} + (-4y + 1)\mathbf{j}$
(D) $(2x + 2)\mathbf{i} + (-4y + 1)\mathbf{j}$

5. What is the divergence of the following vector function?

$$\mathbf{F}(x, y, z) = 2xz\mathbf{i} + e^x y^2\mathbf{j} + x^2 y\mathbf{k}$$

(A) $(2z + e^x + 2x)\mathbf{i} + (2ye^x + x^2)\mathbf{j} + 2x\mathbf{k}$
(B) $2z + 2ye^x$
(C) $2z + 2ye^x + 1$
(D) $2z\mathbf{i} + 2ye^x\mathbf{j} + \mathbf{k}$

SOLUTIONS

1. If c is positive, then $n(\infty)$ is zero, which is contrary to the data given. Therefore, c is less than or equal to zero. If c is zero, then $n(\infty) = a/(1+b) = 100$, which is possible depending on a and b. However, $n(0) = a/(1+b)$ would also equal 100, which is contrary to the given data. Therefore, c is not zero. c must be less than zero.

Because $c < 0$, then $n(\infty) = a$, so $\boxed{a = 100.}$

Applying the condition $t = 0$ gives

$$n(0) = \frac{a}{1+b} = 10$$

Because $a = 100$,

$$n(0) = \frac{100}{1+b}$$
$$100 = (10)(1+b)$$
$$10 = 1+b$$
$$\boxed{b = 9}$$

Substitute the results for a and b into the expression.

$$n(t) = \frac{100}{1 + 9e^{ct}}$$

Take the first derivative.

$$\frac{d}{dt}n(t) = \left(\frac{100}{(1 + 9e^{ct})^2}\right)(-9ce^{ct})$$

Apply the initial condition.

$$\frac{d}{dt}n(0) = \left(\frac{100}{(1 + 9e^{c(0)})^2}\right)\left(-9ce^{c(0)}\right) = 0.5$$
$$\left(\frac{100}{(1 + 9)^2}\right)(-9c) = 0.5$$
$$(1)(-9c) = 0.5$$
$$c = \frac{-0.5}{9}$$
$$= \boxed{-0.056}$$

Substitute the terms a, b, and c into the expression.

$$n(t) = \frac{100}{1 + 9e^{-0.056t}}$$

The answer is (C).

2. Determine the critical points by taking the first derivative of the function and setting it equal to zero.

$$\frac{dy}{dx} = 3x^2 - 18x = 3x(x - 6)$$
$$3x(x - 6) = 0$$
$$x(x - 6) = 0$$

The critical points are located at $x = 0$ and $x = 6$.

Determine the inflection points by setting the second derivative equal to zero. Take the second derivative.

$$\frac{d^2y}{dx^2} = \left(\frac{d}{dx}\right)\left(\frac{dy}{dx}\right) = \frac{d}{dx}(3x^2 - 18x)$$
$$= 6x - 18$$

Set the second derivative equal to zero.

$$\frac{d^2y}{dx^2} = 0 = 6x - 18 = 6(x - 3)$$
$$6(x - 3) = 0$$
$$x - 3 = 0$$
$$x = 3$$

$\boxed{\text{This inflection point is at } x = 3.}$

Determine the local maximum and minimum by substituting the critical points into the expression for the second derivative.

At the critical point $x = 0$,

$$\left.\frac{d^2y}{dx^2}\right|_{x=0} = 6(x - 3) = (6)(0 - 3)$$
$$= -18$$

Because $-18 < 0$, $\boxed{x = 0 \text{ is a local maximum.}}$

At the critical point $x = 6$,

$$\left.\frac{d^2y}{dx^2}\right|_{x=6} = 6(x - 3) = (6)(6 - 3)$$
$$= 18$$

Because $18 > 0$, $\boxed{x = 6 \text{ is a local minimum.}}$

The answer is (A).

3. The derivative of $f(t)$ is

$$f'(t) = 170\frac{d(\cos 377t)}{dt}$$

$$= (170)(-\sin 377t)\left(\frac{d(377t)}{dt}\right)$$

$$= (-170)\sin 377t\left(377\frac{dt}{dt}\right)$$

$$= \boxed{-64{,}090\sin 377t \quad (-64{,}000\sin 377t)}$$

The answer is (D).

4. The gradient of $f(x,y)$ is

$$\nabla f(x,y) = \frac{\partial f(x,y)}{\partial x}\mathbf{i} + \frac{\partial f(x,y)}{\partial y}\mathbf{j}$$

$$= \frac{\partial(x^2 - 2y^2 + 2x + y)}{\partial x}\mathbf{i}$$

$$\quad + \frac{\partial(x^2 - 2y^2 + 2x + y)}{\partial y}\mathbf{j}$$

$$= \boxed{(2x+2)\mathbf{i} + (-4y+1)\mathbf{j}}$$

The answer is (D).

5. The divergence is a scalar, given by

$$\text{div}\,\mathbf{F} = \nabla\cdot\mathbf{F} = \frac{\partial F_x}{\partial x} + \frac{\partial F_y}{\partial y} + \frac{\partial F_z}{\partial z}$$

$$= \frac{\partial(2xz)}{\partial x} + \frac{\partial(e^x y^2)}{\partial y} + \frac{\partial(x^2 y)}{\partial z}$$

$$= 2z + 2e^x y + 0$$

$$= \boxed{2z + 2ye^x}$$

The answer is (B).

10 Integral Calculus

PRACTICE PROBLEMS

1. What is the result of the indefinite integration shown?

$$\int \sin \theta \, d\theta$$

(A) $-\cos \theta + C$
(B) $\cos \theta + C$
(C) $-\sin \theta + C$
(D) $\dfrac{-\sin^2 \theta}{2}$

2. What is the value of the following definite integral?

$$\int_0^{\pi/2} \sin x \, dx$$

(A) -1
(B) 0
(C) 1
(D) $\pi/2$

3. Consider the following square wave function.

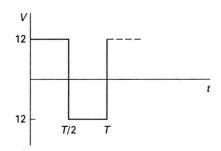

The period of the function is 0.0167 s. What is the value of a_0? That is, what is the value of the first term in a Fourier series representing the function?

(A) -6
(B) 0
(C) $+6$
(D) 12

4. What is the area between the curve shown, $y_1 = e^x$, and the straight line, $y = x + 1$, shown on the interval $x = [0, 2]$?

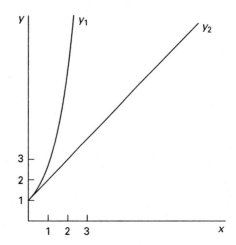

(A) 2.4
(B) 3.4
(C) 5.4
(D) 7.4

5. What is the approximate length of the curve $y = x^2$ from $x = 0$ to $x = 5$?

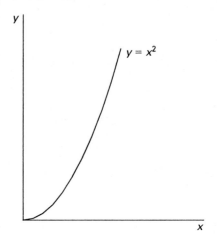

(A) 10
(B) 17
(C) 25
(D) 170

SOLUTIONS

1. The indefinite integral of the sine function is

$$\int \sin \theta \, d\theta = \boxed{-\cos \theta + C}$$

C represents a constant.

The answer is (A).

2. The definite integral is

$$\int_0^{\pi/2} \sin x \, dx = -\cos x \Big|_0^{\pi/2}$$

$$= -\cos \frac{\pi}{2} - (-\cos 0)$$

$$= 0 + 1$$

$$= \boxed{1}$$

The answer is (C).

3. The Fourier series is

$$f(t) = \tfrac{1}{2}a_0 + a_1 \cos \omega t + a_2 \cos 2\omega t + \cdots$$
$$+ b_1 \sin \omega t + b_2 \sin 2\omega t + \cdots$$

Normalizing by dividing all frequencies by ω gives

$$f(t) = \tfrac{1}{2}a_0 + a_1 \cos t + a_2 \cos 2t + \cdots$$
$$+ b_1 \sin t + b_2 \sin 2t + \cdots$$

The first term, $(1/2)a_0$, is the average value of the function. Calculate the value of a_0.

$$a_0 = \frac{2}{T} \int_0^T f(t) dt$$

$$= \frac{2}{T} \left(\int_0^{T/2} 12 dt + \int_{T/2}^T (-12) dt \right)$$

$$= \frac{2}{T} \left(12t \Big|_0^{T/2} - 12t \Big|_{T/2}^T \right)$$

$$= \frac{2}{T} \left(\left(12\frac{T}{2} - (12)(0) \right) + \left((-12T) - (-12)\frac{T}{2} \right) \right)$$

$$= \frac{2}{T} \left(12\frac{T}{2} - 12T + 12\frac{T}{2} \right)$$

$$= \frac{2}{T}(12T - 12T)$$

$$= \frac{2}{T}(0)$$

$$= \boxed{0}$$

The average value, $(1/2)a_0$, equals zero, which can also be determined by inspection of the waveform.

The answer is (B).

4. The area bounded by the two functions is given by the following.

$$A = \int_a^b \big(y_1(x) - y_2(x) \big) dx$$

$$= \int_0^2 \big(e^x - (x+1) \big) dx = \int_0^2 (e^x - x - 1) dx$$

$$= \int_0^2 e^x dx - \int_0^2 x \, dx - \int_0^2 dx$$

$$= e^x \Big|_0^2 - \frac{x^2}{2} \Big|_0^2 - x \Big|_0^2$$

$$= (e^2 - e^0) - \left(\frac{2^2}{2} - \frac{0^2}{2} \right) - (2 - 0)$$

$$= e^2 - 1 - 2 - 2 = e^2 - 5 = 7.389 - 5$$

$$= \boxed{2.389 \quad (2.4)}$$

The answer is (A).

5. The length of the curve may be found from

$$L_{\text{curve}} = \int_a^b \sqrt{1 + \big(f'(x) \big)^2} \, dx \qquad \text{[I]}$$

Because $f(x) = x^2$, the derivative $f'(x) = 2x$. Substitute the derivative function and the given limits into Eq. I.

$$L_{\text{curve}} = \int_0^5 \sqrt{1 + (2x)^2} \, dx$$

$$= \int_0^5 \sqrt{1 + 4x^2} \, dx \qquad \text{[II]}$$

Use a table of integrals to determine the following.

$$\int \sqrt{a^2 + x^2} \, dx = \frac{1}{2} \left(x\sqrt{a^2 + x^2} \right.$$
$$\left. + a^2 \ln \left(x + \sqrt{a^2 + x^2} \right) \right)$$

Put Eq. II into the form of the solution from the table of integrals and solve.

$$L_{\text{curve}} = \int_0^5 \sqrt{1 + 4x^2}\, dx = 2 \int_0^5 \sqrt{\left(\tfrac{1}{2}\right)^2 + x^2}\, dx$$

$$= (2) \left[\left(\tfrac{1}{2}\right) \left(\frac{x\sqrt{\left(\tfrac{1}{2}\right)^2 + x^2} + \left(\tfrac{1}{2}\right)^2}{\times \ln\left(x + \sqrt{\left(\tfrac{1}{2}\right)^2 + x^2}\right)} \right) \Bigg|_0^5 \right]$$

$$= (2) \left[\begin{array}{l} \left(\tfrac{1}{2}\right)\left(\begin{array}{c} 5\sqrt{\left(\tfrac{1}{2}\right)^2 + 5^2} + \left(\tfrac{1}{2}\right)^2 \\ \times \ln\left(5 + \sqrt{\left(\tfrac{1}{2}\right)^2 + 5^2}\right) \end{array} \right) \\[2em] - \left(\tfrac{1}{2}\right)\left(\begin{array}{c} 0\sqrt{\left(\tfrac{1}{2}\right)^2 + 0^2} + \left(\tfrac{1}{2}\right)^2 \\ \times \ln\left(0 + \sqrt{\left(\tfrac{1}{2}\right)^2 + 0^2}\right) \end{array} \right) \end{array} \right]$$

$$= \boxed{25.8 \quad (25)}$$

The answer is (C).

11 Differential Equations

PRACTICE PROBLEMS

1. Solve the following differential equation for y.

$$y'' - 4y' - 12y = 0$$

(A) $A_1 e^{6x} + A_2 e^{-2x}$
(B) $A_1 e^{-6x} + A_2 e^{2x}$
(C) $A_1 e^{6x} + A_2 e^{2x}$
(D) $A_1 e^{-6x} + A_2 e^{-2x}$

2. Solve the following differential equation for y.

$$y' - y = 2xe^{2x} \qquad y(0) = 1$$

(A) $y = 2e^{-2x}(x-1) + 3e^{-x}$
(B) $y = 2e^{2x}(x-1) + 3e^{x}$
(C) $y = -2e^{-2x}(x-1) + 3e^{-x}$
(D) $y = 2e^{2x}(x-1) + 3e^{-x}$

3. The oscillation exhibited by the top story of a certain building in free motion is given by the following differential equation.

$$x'' + 2x' + 2x = 0 \qquad x(0) = 0 \qquad x'(0) = 1$$

(a) What is x as a function of time?
(A) $e^{-2t} \sin t$
(B) $e^{t} \sin t$
(C) $e^{-t} \sin t$
(D) $e^{-t} \sin t + e^{-t} \cos t$

(b) What is the building's fundamental natural frequency of vibration?
(A) $\frac{1}{2}$
(B) 1
(C) $\sqrt{2}$
(D) 2

(c) What is the amplitude of oscillation?
(A) 0.32
(B) 0.54
(C) 1.7
(D) 6.6

(d) What is x as a function of time if a lateral wind load is applied with a form of $\sin t$?

(A) $\frac{2}{5}e^{-t}\cos t + \frac{6}{5}e^{-t}\sin t$

(B) $\frac{2}{5}e^{t}\cos t + \frac{6}{5}e^{t}\sin t$

(C) $\frac{6}{5}e^{-t}\cos t + \frac{2}{5}e^{-t}\sin t - \frac{1}{5}\cos t + \frac{2}{5}\sin t$

(D) $\frac{2}{5}e^{-t}\cos t + \frac{6}{5}e^{-t}\sin t - \frac{2}{5}\cos t + \frac{1}{5}\sin t$

4. (*Time limit: one hour*) A 90 lbm (40 kg) bag of a chemical is accidentally dropped into an aerating lagoon. The chemical is water soluble and nonreacting. The lagoon is 120 ft (35 m) in diameter and filled to a depth of 10 ft (3 m). The aerators circulate and distribute the chemical evenly throughout the lagoon.

Water enters the lagoon at a rate of 30 gal/min (115 L/min). Fully mixed water is pumped into a reservoir at a rate of 30 gal/min (115 L/min).

The established safe concentration of this chemical is 1 ppb (part per billion). Approximately how many days will it take for the concentration of the discharge water to reach this level?
(A) 25 days
(B) 50 days
(C) 100 days
(D) 200 days

5. A tank contains 100 gal (100 L) of brine made by dissolving 60 lbm (60 kg) of salt in pure water. Salt water with a concentration of 1 lbm/gal (1 kg/L) enters the tank at a rate of 2 gal/min (2 L/min). A well-stirred mixture is drawn from the tank at a rate of 3 gal/min (3 L/min). Find the mass of salt in the tank after 1 hr.
(A) 13 lbm (13 kg)
(B) 37 lbm (37 kg)
(C) 43 lbm (43 kg)
(D) 51 lbm (51 kg)

SOLUTIONS

1. Obtain the characteristic equation by replacing each derivative with a polynomial term of equal degree.

$$r^2 - 4r - 12 = 0$$

Factor the characteristic equation.

$$(r - 6)(r + 2) = 0$$

The roots are $r_1 = 6$ and $r_2 = -2$.

Since the roots are real and distinct, the solution is

$$y = A_1 e^{r_1 x} + A_2 e^{r_2 x}$$

$$= \boxed{A_1 e^{6x} + A_2 e^{-2x}}$$

The answer is (A).

2. The equation is a first-order linear differential equation of the form

$$y' + p(x)y = g(x)$$
$$p(x) = -1$$
$$g(x) = 2xe^{2x}$$

The integration factor $u(x)$ is given by

$$u(x) = e^{\left(\int p(x)dx\right)}$$
$$= e^{\left(\int (-1)dx\right)}$$
$$= e^{-x}$$

The closed form of the solution is given by

$$y = \left(\frac{1}{u(x)}\right)\left(\int u(x)g(x)dx + C\right)$$
$$= \left(\frac{1}{e^{-x}}\right)\left(\int (e^{-x})(2xe^{2x})dx + C\right)$$
$$= e^x \left(2(xe^x - e^x) + C\right)$$
$$= e^x \left(2e^x(x - 1) + C\right)$$
$$= 2e^{2x}(x - 1) + Ce^x$$

Apply the initial condition $y(0) = 1$ to obtain the integration constant C.

$$y(0) = 2e^{(2)(0)}(0 - 1) + Ce^0 = 1$$
$$(2)(1)(-1) + C(1) = 1$$
$$-2 + C = 1$$
$$C = 3$$

Substituting in the value for the integration constant C, the solution is

$$y = \boxed{2e^{2x}(x - 1) + 3e^x}$$

The answer is (B).

3. (a) The differential equation is a homogeneous second-order linear differential equation with constant coefficients. Write the characteristic equation.

$$r^2 + 2r + 2 = 0$$

This is a quadratic equation of the form $ar^2 + br + c = 0$, where a equals 1, b equals 2, and c equals 2.

Solve for r.

$$r = \frac{-b \pm \sqrt{b^2 - 4ac}}{2a}$$
$$= \frac{-2 \pm \sqrt{(2)^2 - (4)(1)(2)}}{(2)(1)}$$
$$= \frac{-2 \pm \sqrt{4 - 8}}{2}$$
$$= \frac{-2 \pm \sqrt{-4}}{2}$$
$$= \frac{-2 \pm 2\sqrt{-1}}{2}$$
$$= -1 \pm \sqrt{-1}$$
$$= -1 \pm i$$
$$r_1 = -1 + i \text{ and } r_2 = -1 - i$$

Because the roots are imaginary and of the form $\alpha + i\omega$ and $\alpha - i\omega$, where α equals -1 and ω equals 1, the general form of the solution is given by

$$x(t) = A_1 e^{\alpha t} \cos \omega t + A_2 e^{\alpha t} \sin \omega t$$
$$= A_1 e^{-1t} \cos(1t) + A_2 e^{-1t} \sin(1t)$$
$$= A_1 e^{-t} \cos t + A_2 e^{-t} \sin t$$

Apply the initial conditions $x(0) = 0$ and $x'(0) = 1$ to solve for A_1 and A_2.

First, apply the initial condition $x(0) = 0$.

$$x(t) = A_1 e^0 \cos 0 + A_2 e^0 \sin 0 = 0$$
$$A_1(1)(1) + A_2(1)(0) = 0$$
$$A_1 = 0$$

Substituting, the solution of the differential equation becomes

$$x(t) = A_2 e^{-t} \sin t$$

To apply the second initial condition, take the first derivative.

$$x'(t) = \frac{d}{dt}\left(A_2 e^{-t} \sin t\right)$$
$$= A_2 \frac{d}{dt}\left(e^{-t} \sin t\right)$$
$$= A_2 \left(\sin t \frac{d}{dt}\left(e^{-t}\right) + e^{-t}\frac{d}{dt}\sin t\right)$$
$$= A_2 \left(\sin t\left(-e^{-t}\right) + e^{-t}(\cos t)\right)$$
$$= A_2\left(e^{-t}\right)(-\sin t + \cos t)$$

Apply the initial condition, $x'(0) = 1$.

$$x(0) = A_2\left(e^0\right)(-\sin 0 + \cos 0) = 1$$
$$A_2(1)(0+1) = 1$$
$$A_2 = 1$$

The solution is

$$x(t) = A_2 e^{-t} \sin t$$
$$= (1)e^{-t}\sin t$$
$$= \boxed{e^{-t}\sin t}$$

The answer is (C).

(b) To determine the natural frequency, set the damping term to zero. The equation has the form

$$x'' + 2x = 0$$

This equation has a general solution of the form

$$x(t) = x_0 \cos \omega t + \left(\frac{v_0}{\omega}\right)\sin \omega t$$

ω is the natural frequency.

Given the equation $x''+2x=0$, the characteristic equation is
$$r^2 + 2 = 0$$
$$r = \sqrt{-2}$$
$$= \pm\sqrt{2}i$$

Because the roots are imaginary and of the form $\alpha + i\omega$ and $\alpha - i\omega$, where α equals 0 and ω equals $\sqrt{2}$, the general form of the solution is given by

$$x(t) = A_1 e^{\alpha t}\cos \omega t + A_2 e^{\alpha t}\sin \omega t$$
$$= A_1 e^{0t}\cos \sqrt{2}t + A_2 e^{0t}\sin \sqrt{2}t$$
$$= A_1(1)\cos \sqrt{2}t + A_2(1)\sin \sqrt{2}t$$
$$= A_1 \cos \sqrt{2}t + A_2 \sin \sqrt{2}t$$

Apply the initial conditions $x(0) = 0$ and $x'(0) = 1$ to solve for A_1 and A_2. Applying the initial condition $x(0) = 0$ gives

$$x(0) = A_1 \cos\left((\sqrt{2})(0)\right) + A_2 \sin\left((\sqrt{2})(0)\right) = 0$$
$$A_1 \cos 0 + A_2 \sin 0 = 0$$
$$A_1(1) + A_2(0) = 0$$
$$A_1 = 0$$

Substituting, the solution of the different equation becomes

$$x(t) = A_2 \sin \sqrt{2}t$$

To apply the second initial condition, take the first derivative.

$$x'(t) = \frac{d}{dt}(A_2 \sin \sqrt{2}t)$$
$$= A_2\sqrt{2}\cos \sqrt{2}t$$

Apply the second initial condition, $x'(0) = 1$.

$$x'(0) = A_2\sqrt{2}\cos\left((\sqrt{2})(0)\right) = 1$$
$$A_2\sqrt{2}\cos(0) = 1$$
$$A_2(\sqrt{2})(1) = 1$$
$$A_2\sqrt{2} = 1$$
$$A_2 = \frac{1}{\sqrt{2}} = \frac{\sqrt{2}}{2}$$

Substituting, the undamped solution becomes

$$x(t) = \frac{\sqrt{2}}{2}\sin \sqrt{2}t$$

Therefore, the undamped natural frequency is

$$\boxed{\omega = \sqrt{2}}$$

The answer is (C).

(c) The amplitude of the oscillation is the maximum displacement.

Take the derivative of the solution, $x(t) = e^{-t}\sin t$.

$$x'(t) = \frac{d}{dt}(e^{-t}\sin t)$$
$$= (\sin t)\frac{d}{dt}\left(e^{-t}\right) + (e^{-t})\frac{d}{dt}\sin t$$
$$= \sin t(-e^{-t}) + (e^{-t})\cos t$$
$$= e^{-t}(\cos t - \sin t)$$

The maximum displacement occurs at $x'(t) = 0$.

Because $e^{-t} \neq 0$ except as t approaches infinity,

$$\cos t - \sin t = 0$$
$$\tan t = 1$$
$$t = \tan^{-1} 1$$
$$= 0.785 \text{ rad}$$

At time $t = 0.785$ rad, the displacement is at a maximum. Substitute into the orginal solution to obtain a value for the maximum displacement.

$$x(0.785) = e^{-0.785} \sin 0.785$$
$$= 0.32$$

The amplitude is $\boxed{0.32.}$

The answer is (A).

(d) (An alternative solution using Laplace transforms follows this solution.) The application of a lateral wind load with the form $\sin t$ revises the differential equation to the form

$$x'' + 2x' + 2x = \sin t$$

Express the solution as the sum of the complementary x_c and particular x_p solutions.

$$x(t) = x_c(t) + x_p(t)$$

From part (a),

$$x_c(t) = A_1 e^{-t} \cos t + A_2 e^{-t} \sin t$$

The general form of the particular solution is given by

$$x_p(t) = x^s (A_3 \cos t + A_4 \sin t)$$

Determine the value of s; check to see if the terms of the particular solution solve the homogeneous equation.

Examine the term $A_3 \cos t$.

Take the first derivative.

$$\frac{d}{dt}(A_3 \cos t) = -A_3 \sin t$$

Take the second derivative.

$$\frac{d}{dt}\left(\frac{d}{dt}(A_3 \cos t)\right) = \frac{d}{dt}(-A_3 \sin t)$$
$$= -A_3 \cos t$$

Substitute the terms into the homogeneous equation.

$$x'' + 2x' + 2x = -A_3 \cos t + 2(-A_3 \sin t)$$
$$+ 2(-A_3 \cos t)$$
$$= A_3 \cos t - 2A_3 \sin t$$
$$\neq 0$$

Except for the trival solution $A_3 = 0$, the term $A_3 \cos t$ does not solve the homogeneous equation.

Examine the second term $A_4 \sin t$.

Take the first derivative.

$$\frac{d}{dt}(A_4 \sin t) = A_4 \cos t$$

Take the second derivative.

$$\frac{d}{dt}\left(\frac{d}{dt}(A_4 \sin t)\right) = \frac{d}{dt}(A_4 \cos t)$$
$$= -A_4 \sin t$$

Substitute the terms into the homogeneous equation.

$$x'' + 2x' + 2x = -A_4 \sin t + 2(A_4 \cos t)$$
$$+ 2(A_4 \sin t)$$
$$= A_4 \sin t + 2A_4 \cos t$$
$$\neq 0$$

Except for the trival solution $A_4 = 0$, the term $A_4 \sin t$ does not solve the homogeneous equation.

Neither of the terms satisfies the homogeneous equation $s = 0$; therefore, the particular solution is of the form

$$x_p(t) = A_3 \cos t + A_4 \sin t$$

Use the method of undetermined coefficients to solve for A_3 and A_4. Take the first derivative.

$$x_p'(t) = \frac{d}{dt}(A_3 \cos t + A_4 \sin t)$$
$$= -A_3 \sin t + A_4 \cos t$$

Take the second derivative.

$$x_p''(t) = \frac{d}{dt}\left(\frac{d}{dt}(A_3 \cos t + A_4 \sin t)\right)$$
$$= \frac{d}{dt}(-A_3 \sin t + A_4 \cos t)$$
$$= -A_3 \cos t - A_4 \sin t$$

Substitute the expressions for the derivatives into the differential equation.

$$x'' + 2x' + 2x = (-A_3 \cos t - A_4 \sin t)$$
$$+ 2(-A_3 \sin t + A_4 \cos t)$$
$$+ 2(A_3 \cos t + A_4 \sin t)$$
$$= \sin t$$

Rearranging terms gives

$$(-A_3 + 2A_4 + 2A_3)\cos t$$
$$+(-A_4 - 2A_3 + 2A_4)\sin t = \sin t$$
$$(A_3 + 2A_4)\cos t + (-2A_3 + A_4)\sin t = \sin t$$

Equating coefficients gives

$$A_3 + 2A_4 = 0$$
$$-2A_3 + A_4 = 1$$

Multiplying the first equation by 2 and adding equations gives

$$A_3 + 2A_4 = 0$$
$$+(-2A_3 + A_4) = 1$$
$$\overline{\qquad\qquad\qquad}$$
$$5A_4 = 1 \text{ or } A_4 = \tfrac{1}{5}$$

From the first equation for $A_4 = {}^1\!/_5$, $A_3 + (2)({}^1\!/_5) = 0$ and $A_3 = -{}^2\!/_5$.

Substituting for the coefficients, the particular solution becomes

$$x_p(t) = -\tfrac{2}{5}\cos t + \tfrac{1}{5}\sin t$$

Combining the complementary and particular solutions gives

$$x(t) = x_c(t) + x_p(t)$$
$$= A_1 e^{-t}\cos t + A_2 e^{-t}\sin t - \tfrac{2}{5}\cos t + \tfrac{1}{5}\sin t$$

Apply the initial conditions to solve for the coefficients A_1 and A_2; then apply the first initial condition, $x(0) = 0$.

$$x(t) = A_1 e^0 \cos 0 + A_2 e^0 \sin 0$$
$$- \tfrac{2}{5}\cos 0 + \tfrac{1}{5}\sin 0 = 0$$
$$A_1(1)(1) + A_2(1)(0) + \left(-\tfrac{2}{5}\right)(1) + \left(\tfrac{1}{5}\right)(0) = 0$$
$$A_1 - \tfrac{2}{5} = 0$$
$$A_1 = \tfrac{2}{5}$$

Substituting for A_1, the solution becomes

$$x(t) = \tfrac{2}{5}e^{-t}\cos t + A_2 e^{-t}\sin t - \tfrac{2}{5}\cos t + \tfrac{1}{5}\sin t$$

Take the first derivative.

$$x'(t) = \frac{d}{dt}\left(\tfrac{2}{5}e^{-t}\cos t + A_2 e^{-t}\sin t\right)$$
$$+ \left(\left(-\tfrac{2}{5}\right)\cos t + \tfrac{1}{5}\sin t\right)$$
$$= \left(\tfrac{2}{5}\right)\left(-e^{-t}\cos t - e^{-t}\sin t\right)$$
$$+ A_2\left(-e^{-t}\sin t + e^{-t}\cos t\right)$$
$$+ \left(-\tfrac{2}{5}\right)(-\sin t) + \tfrac{1}{5}\cos t$$

Apply the second initial condition, $x'(0) = 1$.

$$x'(0) = \tfrac{2}{5}\left(-e^0\cos 0 - e^0 \sin 0\right)$$
$$+ A_2\left(-e^0 \sin 0 + e^0 \cos 0\right)$$
$$+ \left(-\tfrac{2}{5}\right)(-\sin 0) + \tfrac{1}{5}\cos 0$$
$$= 1$$

$$\left(\tfrac{2}{5}\right)\left(-(1)(1) - (1)(0)\right)$$
$$+ A_2\left(-(1)(0) + (1)(1)\right) + \left(-\tfrac{2}{5}\right)(0) + \left(\tfrac{1}{5}\right)(1) = 1$$
$$\left(\tfrac{2}{5}\right)(-1) + A_2(1) + \left(\tfrac{1}{5}\right) = 1$$
$$A_2 = \tfrac{6}{5}$$

Substituting for A_2, the solution becomes

$$\boxed{x(t) = \begin{array}{l} \tfrac{2}{5}e^{-t}\cos t + \tfrac{6}{5}e^{-t}\sin t \\ - \tfrac{2}{5}\cos t + \tfrac{1}{5}\sin t \end{array}}$$

The answer is (D).

(d) *Alternate Solution:*

Use the Laplace transform method.

$$x'' + 2x' + 2x = \sin t$$
$$\mathcal{L}(x'') + 2\mathcal{L}(x') + 2\mathcal{L}(x) = \mathcal{L}(\sin t)$$
$$s^2\mathcal{L}(x) - 1 + 2s\mathcal{L}(x) + 2\mathcal{L}(x) = \frac{1}{s^2 + 1}$$
$$\mathcal{L}(x)(s^2 + 2s + 2) - 1 = \frac{1}{s^2 + 1}$$
$$\mathcal{L}(x) = \frac{1}{s^2 + 2s + 2} + \frac{1}{(s^2 + 1)(s^2 + 2s + 2)}$$
$$= \frac{1}{(s+1)^2 + 1} + \frac{1}{(s^2 + 1)(s^2 + 2s + 2)}$$

Use partial fractions to expand the second term.

$$\frac{1}{(s^2 + 1)(s^2 + 2s + 2)} = \frac{A_1 + B_1 s}{s^2 + 1} + \frac{A_2 + B_2 s}{s^2 + 2s + 2}$$

Cross multiply.

$$\frac{A_1 + B_1 s}{s^2 + 1} + \frac{A_2 + B_2 s}{s^2 + 2s + 2}$$

$$= \frac{\begin{array}{l} A_1 s^2 + 2A_1 s + 2A_1 + B_1 s^3 + 2B_1 s^2 \\ + 2B_1 s + A_2 s^2 + A_2 + B_2 s^3 + B_2 s \end{array}}{(s^2 + 1)(s^2 + 2s + 2)}$$

$$= \frac{\begin{array}{l} s^3(B_1 + B_2) + s^2(A_1 + A_2 + 2B_1) \\ + s(2A_1 + 2B_1 + B_2) + 2A_1 + A_2 \end{array}}{(s^2 + 1)(s^2 + 2s + 2)}$$

Compare numerators to obtain the following four simultaneous equations.

$$B_1 + B_2 = 0$$
$$A_1 + A_2 + 2B_1 = 0$$
$$2A_1 + 2B_1 + B_2 = 0$$
$$2A_1 + A_2 = 1$$

Use Cramer's rule to find A_1.

$$A_1 = \frac{\begin{vmatrix} 0 & 0 & 1 & 1 \\ 0 & 1 & 2 & 0 \\ 0 & 0 & 2 & 1 \\ 1 & 1 & 0 & 0 \end{vmatrix}}{\begin{vmatrix} 0 & 0 & 1 & 1 \\ 1 & 1 & 2 & 0 \\ 2 & 0 & 2 & 1 \\ 2 & 1 & 0 & 0 \end{vmatrix}} = \frac{-1}{-5} = \frac{1}{5}$$

The rest of the coefficients are found similarly.

$$A_1 = \tfrac{1}{5}$$
$$A_2 = \tfrac{3}{5}$$
$$B_1 = -\tfrac{2}{5}$$
$$B_2 = \tfrac{2}{5}$$

Then,

$$\mathcal{L}(x) = \frac{1}{(s+1)^2+1} + \frac{\tfrac{1}{5}}{s^2+1} + \frac{-\tfrac{2}{5}s}{s^2+1}$$
$$+ \frac{\tfrac{3}{5}}{s^2+2s+2} + \frac{\tfrac{2}{5}s}{s^2+2s+2}$$

Take the inverse transform.

$$x(t) = \mathcal{L}^{-1}\{\mathcal{L}(x)\}$$
$$= e^{-t}\sin t + \tfrac{1}{5}\sin t - \tfrac{2}{5}\cos t + \tfrac{3}{5}e^{-t}\sin t$$
$$+ \tfrac{2}{5}(e^{-t}\cos t - e^{-t}\sin t)$$
$$= \boxed{\tfrac{2}{5}e^{-t}\cos t + \tfrac{6}{5}e^{-t}\sin t - \tfrac{2}{5}\cos t + \tfrac{1}{5}\sin t}$$

The answer is (D).

4. *Customary U.S. Solution*

The differential equation is given as

$$m'(t) = a(t) - \frac{m(t)o(t)}{V(t)}$$

$a(t) =$ rate of addition of chemical
$m(t) =$ mass of chemical at time t
$o(t) =$ volumetric flow out of the lagoon
$\qquad$ (30 gal/min)
$V(t) =$ volume in the lagoon at time t
$\quad d =$ diameter of the lagoon
$\quad z =$ depth of the lagoon

Water flows into the lagoon at a rate of 30 gal/min, and a water-chemical mix flows out of the lagoon at rate of 30 gal/min. Therefore, the volume of the lagoon at time t is equal to the initial volume.

$$V(t) = \frac{\pi}{4}d^2 z$$
$$= \left(\frac{\pi}{4}\right)(120 \text{ ft})^2(10 \text{ ft})$$
$$= 113{,}097 \text{ ft}^3$$

Use a conversion factor of 7.48 gal/ft³.

$$o(t) = \frac{30 \dfrac{\text{gal}}{\text{min}}}{7.48 \dfrac{\text{gal}}{\text{ft}^3}}$$
$$= 4.01 \text{ ft}^3/\text{min}$$

Substituting into the general form of the differential equation gives

$$m'(t) = a(t) - \frac{m(t)o(t)}{V(t)}$$
$$= 0 \text{ lbm} - m(t)\left(\frac{4.01 \dfrac{\text{ft}^3}{\text{min}}}{113{,}097 \text{ ft}^3}\right)$$
$$= -\left(\frac{3.55\times10^{-5}}{\text{min}}\right)m(t)$$

$$m'(t) + \left(\frac{3.55\times10^{-5}}{\text{min}}\right)m(t) = 0 \text{ lbm}$$

The differential equation of the problem has a characteristic equation.

$$r + \frac{3.55\times10^{-5}}{\text{min}} = 0 \text{ min}^{-1}$$
$$r = -3.55\times10^{-5}/\text{min}$$

The general form of the solution is

$$m(t) = Ae^{rt}$$

Substituting for the root, r, gives

$$m(t) = Ae^{\left(\frac{-3.55\times10^{-5}}{\text{min}}\right)t}$$

Apply the initial condition $m(0) = 90$ lbm at time $t = 0$ min.

$$m(0) = Ae^{\left(\frac{-3.55\times10^{-5}}{\text{min}}\right)(0)} = 90 \text{ lbm}$$
$$Ae^0 = 90 \text{ lbm}$$
$$A = 90 \text{ lbm}$$

Therefore,

$$m(t) = (90 \text{ lbm}) e^{\left(\frac{-3.55 \times 10^{-5}}{\text{min}}\right)t}$$

Solve for t.

$$\frac{m(t)}{90 \text{ lbm}} = e^{\left(\frac{-3.55 \times 10^{-5}}{\text{min}}\right)t}$$

$$\ln \frac{m(t)}{90 \text{ lbm}} = \ln e^{\left(\frac{-3.55 \times 10^{-5}}{\text{min}}\right)t}$$

$$= \frac{-3.55 \times 10^{-5}}{\text{min}} t$$

$$t = \frac{\ln \dfrac{m(t)}{90 \text{ lbm}}}{\dfrac{-3.55 \times 10^{-5}}{\text{min}}}$$

The initial mass of the water in the lagoon is

$$m_i = V\rho$$

$$= (113{,}097 \text{ ft}^3) \left(62.4 \frac{\text{lbm}}{\text{ft}^3}\right)$$

$$= 7.06 \times 10^6 \text{ lbm}$$

The final mass of chemicals is achieved at a concentration of 1 ppb or

$$m_f = \frac{m_i}{1 \times 10^9} = \frac{7.06 \times 10^6 \text{ lbm}}{1 \times 10^9}$$

$$= 7.06 \times 10^{-3} \text{ lbm}$$

Find the time required to achieve a mass of 7.06×10^{-3} lbm.

$$t = \left(\frac{\ln \dfrac{m(t)}{90 \text{ lbm}}}{\dfrac{-3.55 \times 10^{-5}}{\text{min}}}\right) \left(\frac{1 \text{ hr}}{60 \text{ min}}\right) \left(\frac{1 \text{ day}}{24 \text{ hr}}\right)$$

$$= \left(\frac{\ln \dfrac{7.06 \times 10^{-3} \text{ lbm}}{90 \text{ lbm}}}{\dfrac{-3.55 \times 10^{-5}}{\text{min}}}\right) \left(\frac{1 \text{ hr}}{60 \text{ min}}\right) \left(\frac{1 \text{ day}}{24 \text{ hr}}\right)$$

$$= \boxed{185 \text{ days} \quad (200 \text{ days})}$$

The answer is (D).

SI Solution

The differential equation is given as

$$m'(t) = a(t) - \frac{m(t)o(t)}{V(t)}$$

$a(t) = $ rate of addition of chemical

$m(t) = $ mass of chemical at time t

$o(t) = $ volumetric flow out of the lagoon

(115 L/min)

$V(t) = $ volume in the lagoon at time t

$d = $ diameter of the lagoon

$z = $ depth of the lagoon

Water flows into the lagoon at a rate of 115 L/min, and a water-chemical mix flows out of the lagoon at a rate of 115 L/min. Therefore, the volume of the lagoon at time t is equal to the initial volume.

$$V(t) = \frac{\pi}{4}d^2 z$$

$$= \left(\frac{\pi}{4}\right)(35 \text{ m})^2(3 \text{ m})$$

$$= 2886 \text{ m}^3$$

Using a conversion factor of 1 m³/1000 L gives

$$o(t) = \left(115 \frac{\text{L}}{\text{min}}\right) \left(\frac{1 \text{ m}^3}{1000 \text{ L}}\right)$$

$$= 0.115 \text{ m}^3/\text{min}$$

Substitute into the general form of the differential equation.

$$m'(t) = a(t) - \frac{m(t)o(t)}{V(t)}$$

$$= 0 \text{ lbm} - m(t) \left(\frac{0.115 \dfrac{\text{m}^3}{\text{min}}}{2886 \text{ m}^3}\right)$$

$$= -\left(\frac{3.985 \times 10^{-5}}{\text{min}}\right) m(t)$$

$$m'(t) + \left(\frac{3.985 \times 10^{-5}}{\text{min}}\right) m(t) = 0 \text{ kg}$$

The differential equation of the problem has the following characteristic equation.

$$r + \frac{3.985 \times 10^{-5}}{\text{min}} = 0 \text{ min}^{-1}$$

$$r = -3.985 \times 10^{-5}/\text{min}$$

The general form of the solution is

$$m(t) = Ae^{rt}$$

Substituting in for the root, r, gives

$$m(t) = Ae^{\left(\frac{-3.985 \times 10^{-5}}{\text{min}}\right)t}$$

Apply the initial condition $m(0) = 40$ kg at time $t = 0$ min.

$$m(0) = Ae^{\left(\frac{-3.985 \times 10^{-5}}{\text{min}}\right)\leftarrow 0 \leftarrow} = 40 \text{ kg}$$

$$Ae^0 = 40 \text{ kg}$$

$$A = 40 \text{ kg}$$

Therefore,

$$m(t) = (40 \text{ kg})e^{\left(\frac{-3.985 \times 10^{-5}}{\text{min}}\right)t}$$

Solve for t.

$$\frac{m(t)}{40 \text{ kg}} = e^{\left(\frac{-3.985 \times 10^{-5}}{\text{min}}\right)t}$$

$$\ln\left(\frac{m(t)}{40 \text{ kg}}\right) = \ln e^{\left(\frac{-3.985 \times 10^{-5}}{\text{min}}\right)t}$$

$$= \left(\frac{-3.985 \times 10^{-5}}{\text{min}}\right)t$$

$$t = \frac{\ln\dfrac{m(t)}{40 \text{ kg}}}{\dfrac{-3.985 \times 10^{-5}}{\text{min}}}$$

The initial mass of water in the lagoon is

$$m_i = V\rho$$

$$= (2886 \text{ m}^3)\left(1000 \ \frac{\text{kg}}{\text{m}^3}\right)$$

$$= 2.886 \times 10^6 \text{ kg}$$

The final mass of chemicals is achieved at a concentration of 1 ppb or

$$m_f = \frac{2.886 \times 10^6 \text{ kg}}{1 \times 10^9}$$

$$= 2.886 \times 10^{-3} \text{ kg}$$

Find the time required to achieve a mass of 2.886 $\times 10^{-3}$ kg.

$$t = \left(\frac{\ln\dfrac{m(t)}{40 \text{ kg}}}{\dfrac{-3.985 \times 10^{-5}}{\text{min}}}\right)\left(\frac{1 \text{ h}}{60 \text{ min}}\right)\left(\frac{1 \text{ d}}{24 \text{ h}}\right)$$

$$= \left(\frac{\ln\left(\dfrac{2.886 \times 10^{-3} \text{ kg}}{40 \text{ kg}}\right)}{\dfrac{-3.985 \times 10^{-5}}{\text{min}}}\right)\left(\frac{1 \text{ h}}{60 \text{ min}}\right)\left(\frac{1 \text{ d}}{24 \text{ h}}\right)$$

$$= \boxed{166 \text{ d} \quad (200 \text{ d})}$$

The answer is (D).

5. Let

$$m(t) = \text{mass of salt in tank at time } t$$

$$m_0 = 60 \text{ mass units}$$

$$m'(t) = \text{rate at which salt content is changing}$$

Two mass units of salt enter each minute, and three volumes leave each minute. The amount of salt leaving each minute is

$$\left(3 \ \frac{\text{vol}}{\text{min}}\right)\left(\text{concentration in } \frac{\text{mass}}{\text{vol}}\right)$$

$$= \left(3 \ \frac{\text{vol}}{\text{min}}\right)\left(\frac{\text{salt content}}{\text{volume}}\right)$$

$$= \left(3 \ \frac{\text{vol}}{\text{min}}\right)\left(\frac{m(t)}{100 - t}\right)$$

$$m'(t) = 2 - (3)\left(\frac{m(t)}{100 - t}\right)$$

Alternately,

$$m'(t) + \frac{3m(t)}{100 - t} = 2 \text{ mass/min}$$

This is a first-order linear differential equation. The integrating factor is

$$m = e^{3\int \frac{dt}{100-t}}$$

$$= e^{(3)\left(-\ln(100-t)\right)}$$

$$= (100 - t)^{-3}$$

$$m(t) = (100 - t)^3\left(2\int \frac{dt}{(100 - t)^3} + k\right)$$

$$= 100 - t + k(100 - t)^3$$

But $m = 60$ mass units at time $t = 0$ min, so k is -0.00004.

$$m(t) = 100 - t - (0.00004)(100 - t)^3$$

At time $t = 60$ min,

$$m = 100 - 60 \text{ min} - (0.00004)(100 - 60 \text{ min})^3$$

$$= \boxed{37.44 \text{ mass units} \quad (37 \text{ mass units})}$$

The answer is (B).

12 Probability and Statistical Analysis of Data

PRACTICE PROBLEMS

1. Four military recruits whose respective shoe sizes are 7, 8, 9, and 10 report to the supply clerk to be issued boots. The supply clerk selects one pair of boots in each of the four required sizes and hands them at random to the recruits.

(a) What is the probability that all recruits will receive boots of an incorrect size?
- (A) 0.25
- (B) 0.38
- (C) 0.45
- (D) 0.61

(b) What is the probability that exactly three recruits will receive boots of the correct size?
- (A) 0.0
- (B) 0.063
- (C) 0.17
- (D) 0.25

2. The time taken by a toll taker to collect the toll from vehicles crossing a bridge is an exponential distribution with a mean of 23 sec. What is the probability that a random vehicle will be processed in 25 sec or more (i.e., will take longer than 25 sec)?
- (A) 0.17
- (B) 0.25
- (C) 0.34
- (D) 0.52

3. The number of cars entering a toll plaza on a bridge during the hour after midnight follows a Poisson distribution with a mean of 20.

(a) What is the probability that 17 cars will pass through the toll plaza during that hour on any given night?
- (A) 0.076
- (B) 0.12
- (C) 0.16
- (D) 0.23

(b) What is the probability that three or fewer cars will pass through the toll plaza at that hour on any given night?
- (A) 0.0000032
- (B) 0.0019
- (C) 0.079
- (D) 0.11

4. A mechanical component exhibits a negative exponential failure distribution with a mean time to failure, MTTF, of 1000 hr. What is the maximum operating time such that the reliability remains above 99%?
- (A) 3.3 hr
- (B) 5.6 hr
- (C) 8.1 hr
- (D) 10 hr

5. A survey field crew measures one leg of a traverse four times. The following results are obtained.

repetition	measurement	direction
1	1249.529	forward
2	1249.494	backward
3	1249.384	forward
4	1249.348	backward

The crew chief is under orders to obtain readings with confidence limits of 90%.

(a) Which readings are acceptable?
- (A) No readings are acceptable.
- (B) Two readings are acceptable.
- (C) Three readings are acceptable.
- (D) All four readings are acceptable.

(b) Which readings are NOT acceptable?
- (A) No readings are unacceptable.
- (B) One reading is unacceptable.
- (C) Two readings are unacceptable.
- (D) All four readings are unacceptable.

(c) Explain how to determine which readings are NOT acceptable.
  (A) Readings inside the 90% confidence limits are unacceptable.
  (B) Readings outside the 90% confidence limits are unacceptable.
  (C) Readings outside the upper 90% confidence limit are unacceptable.
  (D) Readings outside the lower 90% confidence limit are unacceptable.

(d) What is the most probable value of the distance?
  (A) 1249.399
  (B) 1249.410
  (C) 1249.439
  (D) 1249.452

(e) What is the error in the most probable value (at 90% confidence)?
  (A) 0.080
  (B) 0.11
  (C) 0.14
  (D) 0.19

(f) If the distance is one side of a square traverse whose sides are all equal, what is the most probable closure error?
  (A) 0.14
  (B) 0.20
  (C) 0.28
  (D) 0.35

(g) What is the probable error of part (f) expressed as a fraction?
  (A) 1:17,600
  (B) 1:14,200
  (C) 1:12,500
  (D) 1:10,900

(h) What is the order of accuracy of the closure?
  (A) first order
  (B) second order
  (C) third order
  (D) fourth order

(i) Define accuracy, and distinguish it from precision.
  (A) If an experiment can be repeated with identical results, the results are considered accurate.
  (B) If an experiment has a small bias, the results are considered precise.
  (C) If an experiment is precise, it cannot also be accurate.
  (D) If an experiment is unaffected by experimental error, the results are accurate.

(j) Give an example of systematic error.
  (A) measuring a river's depth as a motorized ski boat passes by
  (B) using a steel tape that is too short to measure consecutive distances
  (C) locating magnetic north near a large iron ore deposit along an overland route
  (D) determining local wastewater BOD after a toxic spill

6. Prior to the use of radar speed control on a road, California law requires a statistical analysis of the average speed driven there by motorists. The following speeds (all in mph) were observed in a random sample of 40 cars.

44, 48, 26, 25, 20, 43, 40, 42, 29, 39, 23, 26, 24, 47, 45, 28, 29, 41, 38, 36, 27, 44, 42, 43, 29, 37, 34, 31, 33, 30, 42, 43, 28, 41, 29, 36, 35, 30, 32, 31

(a) Tabulate the frequency distribution of the data.

(b) Draw the frequency histogram.

(c) Draw the frequency polygon.

(d) Tabulate the cumulative frequency distribution.

(e) Draw the cumulative frequency graph.

(f) What is the upper quartile speed?
  (A) 30 mph
  (B) 35 mph
  (C) 40 mph
  (D) 45 mph

(g) What is the median speed?
  (A) 31 mph
  (B) 33 mph
  (C) 35 mph
  (D) 37 mph

(h) What is the standard deviation of the sample data?
  (A) 2.1 mph
  (B) 6.1 mph
  (C) 6.8 mph
  (D) 7.4 mph

(i) What is the sample standard deviation?
  (A) 7.5 mph
  (B) 18 mph
  (C) 35 mph
  (D) 56 mph

(j) What is most nearly the sample variance?
  (A) 56 mi^2/hr^2
  (B) 320 mi^2/hr^2
  (C) 1230 mi^2/hr^2
  (D) 3140 mi^2/hr^2

7. A spot speed study is conducted for a stretch of roadway. During a normal day, the speeds were found to be normally distributed with a mean of 46 mph and a standard deviation of 3 mph.

(a) What is the 50th percentile speed?
  (A) 39 mph
  (B) 43 mph
  (C) 46 mph
  (D) 49 mph

(b) What is the 85th percentile speed?
  (A) 47.1 mph
  (B) 48.3 mph
  (C) 49.1 mph
  (D) 52.7 mph

(c) What speed is two standard deviations above the mean?
  (A) 47.2 mph
  (B) 49.3 mph
  (C) 51.1 mph
  (D) 52.0 mph

(d) The daily average speeds for the same stretch of roadway on consecutive normal days were determined by sampling 25 vehicles each day. What speed is two standard deviations above the mean?
  (A) 46.6 mph
  (B) 47.2 mph
  (C) 52.0 mph
  (D) 54.7 mph

8. The diameters of bolt holes drilled in structural steel members are normally distributed with a mean of 0.502 in and a standard deviation of 0.005 in. Holes are out of specification if their diameters are less than 0.497 in or more than 0.507 in.

(a) What is the probability that a hole chosen at random will be out of specification?
  (A) 0.16
  (B) 0.22
  (C) 0.32
  (D) 0.68

(b) What is the probability that 2 holes out of a sample of 15 will be out of specification?
  (A) 0.074
  (B) 0.12
  (C) 0.15
  (D) 0.32

9. 100 bearings were tested to failure. The average life was 1520 hr, and the standard deviation was 120 hr. The manufacturer claims a 1600 hr life. Evaluate using confidence limits of 95% and 99%.
  (A) The claim is accurate at both 95% and 99% confidence.
  (B) The claim is inaccurate only at 95%.
  (C) The claim is inaccurate only at 99%.
  (D) The claim is inaccurate at both 95% and 99% confidence.

10. (a) Find the best equation for a line passing through the points given. (b) Find the correlation coefficient.

x	y
400	370
800	780
1250	1210
1600	1560
2000	1980
2500	2450
4000	3950

11. Find the best equation for a line passing through the points given.

s	t
20	43
18	141
16	385
14	1099

12. The number of vehicles lining up behind a flashing railroad crossing has been observed for five trains of different lengths, as given. What is the mathematical formula that relates the two variables?

no. of cars in train	no. of vehicles
2	14.8
5	18.0
8	20.4
12	23.0
27	29.9

13. The following yield data are obtained from five identical treatment plants. (a) Develop a mathematical equation to correlate the yield and average temperature. (b) What is the correlation coefficient?

treatment plant	average temperature (T)	average yield (Y)
1	207.1	92.30
2	210.3	92.58
3	200.4	91.56
4	201.1	91.63
5	203.4	91.83

14. The following data are obtained from a soil compaction test. What is the mathematical formula that relates the two variables?

x	y
−1	0
0	1
1	1.4
2	1.7
3	2
4	2.2
5	2.4
6	2.6
7	2.8
8	3

15. Two resistances, the meter resistance and a shunt resistor, are connected in parallel in an ammeter. Most of the current passing through the meter goes through the shunt resistor. In order to determine the accuracy of the resistance of shunt resistors being manufactured for a line of ammeters, a manufacturer tests a sample of 100 shunt resistors. The numbers of shunt resistors with the resistance indicated (to the nearest hundredth of an ohm) are as follows.

0.200 Ω, 1; 0.210 Ω, 3; 0.220 Ω, 5; 0.230 Ω, 10; 0.240 Ω, 17; 0.250 Ω, 40; 0.260 Ω, 13; 0.270 Ω, 6; 0.280 Ω, 3; 0.290 Ω, 2

(a) What is the mean resistance?
- (A) 0.235 Ω
- (B) 0.247 Ω
- (C) 0.251 Ω
- (D) 0.259 Ω

(b) What is the sample standard deviation?
- (A) 0.00030
- (B) 0.010
- (C) 0.016
- (D) 0.24

(c) What is the median resistance?
- (A) 0.221 Ω
- (B) 0.244 Ω
- (C) 0.252 Ω
- (D) 0.259 Ω

(d) What is most nearly the sample variance?
- (A) 0.00027 Ω^2
- (B) 0.0083 Ω^2
- (C) 0.011 Ω^2
- (D) 0.016 Ω^2

SOLUTIONS

1. (a) There are 4! = 24 different possible outcomes. By enumeration, there are 9 completely wrong combinations.

$$p\{\text{all wrong}\} = \frac{9}{24} = \boxed{0.375 \quad (0.38)}$$

The answer is (B).

(b) If three recruits get the correct size, the fourth recruit will also, because there will be only one pair remaining.

$$p\{\text{exactly 3}\} = \boxed{0}$$

The answer is (A).

2. For an exponential distribution function, the mean is given as

$$\mu = \frac{1}{\lambda}$$

For a mean of 23,

$$\mu = 23 = \frac{1}{\lambda}$$
$$\lambda = 0.0435$$

For an exponential distribution function,

$$p\{X < x\} = F(x) = 1 - e^{-\lambda x}$$
$$p\{x > X\} = 1 - p\{X < x\}$$
$$= 1 - F(x)$$
$$= 1 - \left(1 - e^{-\lambda x}\right)$$
$$= e^{-\lambda x}$$

The probability of a random vehicle being processed in 25 sec or more is

$$p\{x > 25\} = e^{-(0.0435)(25)}$$
$$= e^{-1.0875}$$
$$= \boxed{0.337 \quad (0.34)}$$

The answer is (C).

3. (a) The distribution is a Poisson distribution with an average of $\lambda = 20$. The probability for a Poisson distribution is

$$p\{x\} = f(x) = \frac{e^{-\lambda}\lambda^x}{x!}$$

The probability of 17 cars is

$$p\{x = 17\} = f(17) = \frac{e^{-20} \times 20^{17}}{17!}$$
$$= \boxed{0.076 \quad (7.6\%)}$$

The answer is (A).

(b) The probability of three or fewer cars is

$$p\{x \le 3\} = p\{x = 0\} + p\{x = 1\} + p\{x = 2\}$$
$$+ p\{x = 3\}$$

$$= f(0) + f(1) + f(2) + f(3)$$

$$= \frac{e^{-20} \times 20^0}{0!} + \frac{e^{-20} \times 20^1}{1!}$$
$$+ \frac{e^{-20} \times 20^2}{2!} + \frac{e^{-20} \times 20^3}{3!}$$

$$= 2 \times 10^{-9} + 4.1 \times 10^{-8}$$
$$+ 4.12 \times 10^{-7} + 2.75 \times 10^{-6}$$

$$= 3.2 \times 10^{-6} \quad (3.2 \times 10^{-4}\%)$$

$$= \boxed{0.0000032}$$

The answer is (A).

4. To simplify the notation,

$$\lambda = \frac{1}{\text{MTTF}}$$

$$\frac{1}{1000 \text{ hr}} = 0.001 \text{ hr}^{-1}$$

The reliability function is

$$R\{t\} = e^{-\lambda t/\text{hr}} = e^{-0.001t/\text{hr}}$$

Because the reliability is greater than 99%,

$$e^{-0.001t/\text{hr}} > 0.99$$
$$\ln(e^{-0.001t/\text{hr}}) > \ln 0.99$$
$$\frac{-0.001t}{\text{hr}} > \ln 0.99$$
$$t < (-1000 \text{ hr}) \ln 0.99$$

$$\boxed{t < 10.05 \text{ hr} \quad (10 \text{ hr})}$$

The maximum operating time such that the reliability remains above 99% is 10.05 hr.

The answer is (D).

5. Find the average.

$$\bar{x} = \frac{\sum x_i}{n}$$

$$= \frac{1249.529 + 1249.494 + 1249.384 + 1249.348}{4}$$

$$= 1249.439$$

Because the sample population is small, use the sample standard deviation.

$$s = \sqrt{\frac{\sum (x_i - \bar{x})^2}{n - 1}}$$

$$= \sqrt{\frac{\begin{array}{c}(1249.529 - 1249.439)^2 + (1249.494 - 1249.439)^2 \\ + (1249.384 - 1249.439)^2 + (1249.348 - 1249.439)^2\end{array}}{4 - 1}}$$

$$= 0.08647$$

From the standard deviation table, a 90% confidence limit falls within $1.645s$ of $\bar{x}$.

$$1249.439 \pm (1.645)(0.08647) = 1249.439 \pm 0.142$$

Therefore, (1249.297, 1249.581) is the 90% confidence range.

(a) $\boxed{\text{By observation, all the readings fall within the 90\% confidence range.}}$

The answer is (D).

(b) $\boxed{\text{No readings are unacceptable.}}$

The answer is (A).

(c) $\boxed{\text{Readings outside the 90\% confidence limits are unacceptable.}}$

The answer is (B).

(d) The unbiased estimate of the most probable distance is $\boxed{1249.439.}$

The answer is (C).

(e) The error for the 90% confidence range is $\boxed{0.14.}$

The answer is (C).

(f) If the surveying crew places a marker, measures a distance x, places a second marker, and then measures the same distance x back to the original marker, the ending point should coincide with the original marker. If, due to measurement errors, the ending and starting points do not coincide, the difference is the closure error.

In this example, the survey crew moves around the four sides of a square, so there are two measurements in the x-direction and two measurements in the y-direction. If the errors E_1 and E_2 are known for two measurements, x_1 and x_2, the error associated with the sum or difference $x_1 \pm x_2$ is

$$E\{x_1 \pm x_2\} = \sqrt{E_1^2 + E_2^2}$$

In this case, the error in the x-direction is

$$E_x = \sqrt{(0.1422)^2 + (0.1422)^2}$$
$$= 0.2011$$

The error in the y-direction is calculated the same way and is also 0.2011. E_x and E_y are combined by the Pythagorean theorem to yield

$$E_{\text{closure}} = \sqrt{(0.2011)^2 + (0.2011)^2}$$

$$= \boxed{0.2844 \quad (0.28)}$$

The answer is (C).

(g) In surveying, error may be expressed as a fraction of one or more legs of the traverse. Assume that the total of all four legs is to be used as the basis.

$$E = \frac{E_{\text{closure}}}{\sum_i x_i} = \frac{0.2844}{(4)(1249)} = \boxed{\frac{1}{17{,}567} \quad (1{:}17{,}600)}$$

The answer is (A).

(h) In surveying, a class 1 third-order error is smaller than 1/10,000. The error of 1/17,567 is smaller than the third-order error; therefore, the error is within the third-order accuracy.

The answer is (C).

(i) An experiment is accurate if it is unchanged by experimental error. Precision is concerned with the repeatability of the experimental results. If an experiment is repeated with identical results, the experiment is said to be precise. However, it is possible to have a highly precise experiment with a large bias.

The answer is (D).

(j) A systematic error is one that is always present and is unchanged from sample to sample. For example, a steel tape that is 0.02 ft short introduces a systematic error.

The answer is (B).

6. (a) and (d) Tabulate the frequency distribution data.

(Note that the lowest speed is 20 mph and the highest speed is 48 mph; therefore, the range is 28 mph. Choose 10 cells with a width of 3 mph.)

midpoint	interval (mph)	frequency	cumulative frequency	cumulative percent
21	20–22	1	1	3
24	23–25	3	4	10
27	26–28	5	9	23
30	29–31	8	17	43
33	32–34	3	20	50
36	35–37	4	24	60
39	38–40	3	27	68
42	41–43	8	35	88
45	44–46	3	38	95
48	47–49	2	40	100

(b)

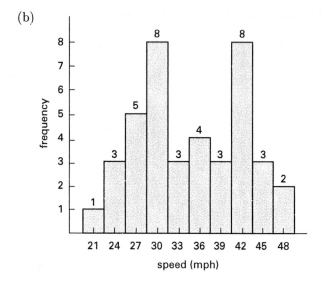

(c)

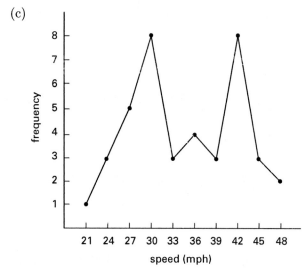

(e)

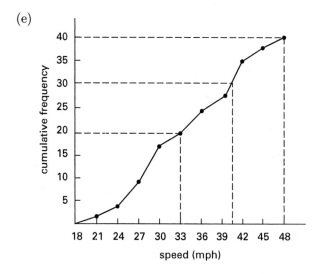

(f) From the cumulative frequency graph in part (e), the upper quartile speed occurs at 30 cars or 75%, which corresponds to approximately 40 mph.

The answer is (C).

(g) The mode occurs at two speeds (the frequency at each of the two speeds is 8), which are 30 mph and 42 mph.

The median occurs at 50% or 20 cars and, from the cumulative frequency chart, corresponds to 33 mph.

$$\sum x_i = 1390 \text{ mi/hr}$$
$$n = 40$$

The answer is (B).

(h) The standard deviation of the sample data is given as

$$\sigma = \sqrt{\frac{\sum x^2}{n} - \mu^2}$$
$$\sum x^2 = 50{,}496 \text{ mi}^2/\text{hr}^2$$

The mean is computed as

$$\bar{x} = \frac{\sum x_i}{n}$$
$$= \frac{1390 \ \dfrac{\text{mi}}{\text{hr}}}{40}$$
$$= \boxed{34.75 \text{ mph} \quad (35 \text{ mph})}$$

Use the sample mean as an unbiased estimator of the population mean, μ.

$$\sigma = \sqrt{\frac{\sum x^2}{n} - \mu^2}$$
$$= \sqrt{\frac{50{,}496 \ \dfrac{\text{mi}^2}{\text{hr}^2}}{40} - \left(34.75 \ \dfrac{\text{mi}}{\text{hr}}\right)^2}$$
$$= \boxed{7.405 \text{ mph} \quad (7.4 \text{ mph})}$$

The answer is (D).

(i) The sample standard deviation is

$$s = \sqrt{\frac{\sum x^2 - \dfrac{(\sum x)^2}{n}}{n-1}}$$
$$= \sqrt{\frac{50{,}496 \ \dfrac{\text{mi}^2}{\text{hr}^2} - \dfrac{\left(1390 \ \dfrac{\text{mi}}{\text{hr}}\right)^2}{40}}{40-1}}$$
$$= \boxed{7.500 \text{ mph} \quad (7.5 \text{ mph})}$$

The answer is (A).

(j) The sample variance is given by the square of the sample standard deviation.

$$s^2 = \left(7.500 \ \frac{\text{mi}}{\text{hr}}\right)^2$$
$$= \boxed{56.25 \text{ mi}^2/\text{hr}^2 \quad (56 \text{ mi}^2/\text{hr}^2)}$$

The answer is (A).

7. (a) The 50th percentile speed is the mean speed,

$$\boxed{46 \text{ mph}}$$

The answer is (C).

(b) The 85th percentile speed is the speed that is exceeded by only 15% of the measurements. Because this is a normal distribution, App. 12.A can be used. 15% in the upper tail corresponds to 35% between the mean and the 85th percentile. This occurs at approximately 1.04σ. The 85th percentile speed is

$$x_{85\%} = \mu + 1.04\sigma$$
$$= 46 \text{ mph} + (1.04)(3 \text{ mph})$$
$$= \boxed{49.12 \text{ mph} \quad (49.1 \text{ mph})}$$

The answer is (C).

(c) The upper two standard deviation speed, 2σ, is

$$x_{2\sigma} = \mu + 2\sigma$$
$$= 46 \text{ mph} + (2)(3 \text{ mph}) = \boxed{52 \text{ mph}}$$

The answer is (D).

(d) According to the central limit theorem, the mean of the average speeds is the same as the distribution mean, and, for a sample size of K measurements, the standard deviation of sample means is

$$\sigma_{\bar{x}} = \frac{\sigma_x}{\sqrt{K}}$$
$$= \frac{3 \text{ mph}}{\sqrt{25}} = 0.6 \text{ mph}$$
$$\bar{x}_{2\sigma} = \mu + 2\sigma_{\bar{x}}$$
$$= 46 \text{ mph} + (2)(0.6 \text{ mph}) = \boxed{47.2 \text{ mph}}$$

The answer is (B).

8. (a) From Eq. 12.43,

$$z = \frac{x_o - \mu}{\sigma}$$
$$z_{\text{upper}} = \frac{0.507 \text{ in} - 0.502 \text{ in}}{0.005 \text{ in}} = +1$$

From App. 12.A, the area outside $z = +1$ is

$$0.5 - 0.3413 = 0.1587$$

Because these are symmetrical limits, $z_{lower} = -1$.

$$\text{total fraction defective} = (2)(0.1587)$$

$$= \boxed{0.3174 \quad (0.32)}$$

The answer is (C).

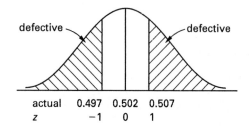

defective					defective
actual	0.497	0.502	0.507		
z		−1	0	1	

(b) This is a binomial problem.

$$p = p\{\text{defective}\} = 0.3174$$

$$q = 1 - p = 0.6826$$

From Eq. 12.28,

$$p\{x\} = f(x) = \binom{n}{x} \hat{p}^x - \hat{q}^{n-x}$$

$$f(2) = \binom{15}{2} (0.3174)^2 (0.6826)^{13}$$

$$= \left(\frac{15!}{13!2!}\right) (0.3174)^2 (0.6826)^{13}$$

$$= \boxed{0.0739 \quad (0.074)}$$

The answer is (A).

9. This is a typical hypothesis test of two sample population means. The two populations are the original population the manufacturer used to determine the 1600 hr average life value and the new population the sample was taken from. The mean ($\bar{x} = 1520$ hr) of the sample and its standard deviation ($s = 120$ hr) are known, but the mean and standard deviation of a population of average lifetimes are unknown.

Assume that the average lifetime population mean and the sample mean are identical.

$$\bar{x} = \mu = 1520 \text{ hr}$$

The standard deviation of the average lifetime population is

$$\sigma_{\bar{x}} = \frac{s}{\sqrt{n}} = \frac{120 \text{ hr}}{\sqrt{100}} = 12 \text{ hr}$$

The manufacturer can be reasonably sure that the claim of a 1600 hr average life is justified if the average test life is near 1600 hr. "Reasonably sure" must be evaluated based on acceptable probability of being incorrect. If the manufacturer is willing to be wrong with a 5% probability, then a 95% confidence level is required.

Since the direction of bias is known, a one-tailed test is required. To determine if the mean has shifted downward, test the hypothesis that 1600 hr is within the 95% limit of a distribution with a mean of 1520 hr and a standard deviation of 12 hr. From a standard normal table, 5% of a standard normal distribution is outside of z equals 1.645.

$$\text{confidence limit} = \bar{x} + z\sigma_x$$

Therefore,

$$95\% \text{ confidence limit} = 1520 \text{ hr} + (1.645)(12 \text{ hr})$$

$$= 1540 \text{ hr}$$

The manufacturer can be 95% certain that the average lifetime of the bearings is less than 1600 hr.

If the manufacturer is willing to be wrong with a probability of only 1%, then a 99% confidence limit is required. From the normal table, $z = 2.33$, and the 99% confidence limit is

$$99\% \text{ confidence limit} = 1520 \text{ hr} + (2.33)(12 \text{ hr})$$

$$= 1548 \text{ hr}$$

> The manufacturer can be 99% certain that the average bearing life is less than 1600 hr.

The answer is (D).

10. (a) Plot the data points to determine if the relationship is linear.

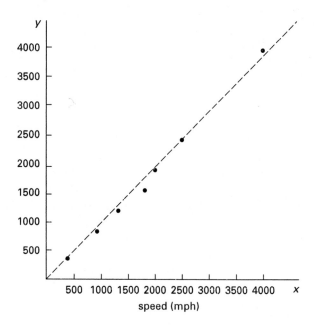

The data appear to be essentially linear. The slope, m, and the y-intercept, b, can be determined using linear regression.

The individual terms are

$$n = 7$$

$$\sum x_i = 400 + 800 + 1250 + 1600 + 2000 + 2500 + 4000$$
$$= 12{,}550$$

$$\left(\sum x_i\right)^2 = (12{,}550)^2$$
$$= 1.575 \times 10^8$$

$$\overline{x} = \frac{\sum x_i}{n}$$
$$= \frac{12{,}550}{7}$$
$$= 1792.9$$

$$\sum x_i^2 = (400)^2 + (800)^2 + (1250)^2 + (1600)^2 + (2000)^2 + (2500)^2 + (4000)^2$$
$$= 3.117 \times 10^7$$

Similarly,

$$\sum y_i = 370 + 780 + 1210 + 1560 + 1980 + 2450 + 3950$$
$$= 12{,}300$$

$$\left(\sum y_i\right)^2 = (12{,}300)^2 = 1.513 \times 10^8$$

$$\overline{y} = \frac{\sum y_i}{n} = \frac{12{,}300}{7} = 1757.1$$

$$\sum y_i^2 = (370)^2 + (780)^2 + (1210)^2 + (1560)^2 + (1980)^2 + (2450)^2 + (3950)^2$$
$$= 3.017 \times 10^7$$

Also,

$$\sum x_i y_i = (400)(370) + (800)(780) + (1250)(1210) + (1600)(1560) + (2000)(1980) + (2500)(2450) + (4000)(3950)$$
$$= 3.067 \times 10^7$$

The slope is

$$m = \frac{n \sum x_i y_i - \sum x_i \sum y_i}{n \sum x_i^2 - \left(\sum x_i\right)^2}$$
$$= \frac{(7)(3.067 \times 10^7) - (12{,}550)(12{,}300)}{(7)(3.117 \times 10^7) - (12{,}550)^2}$$
$$= 0.994$$

The y-intercept is

$$b = \overline{y} - m\overline{x}$$
$$= 1757.1 - (0.994)(1792.9)$$
$$= -25.0$$

The least squares equation of the line is

$$y = mx + b$$
$$= \boxed{0.994x - 25.0}$$

(b) The correlation coefficient is

$$r = \frac{n \sum(x_i y_i) - \left(\sum x_i\right)\left(\sum y_i\right)}{\sqrt{\left(n \sum x_i^2 - \left(\sum x_i\right)^2\right)\left(n \sum y_i^2 - \left(\sum y_i\right)^2\right)}}$$

$$= \frac{(7)(3.067 \times 10^7) - (12{,}500)(12{,}300)}{\sqrt{\begin{array}{c}\left((7)(3.117 \times 10^7) - (12{,}500)^2\right) \\ \times \left((7)(3.017 \times 10^7) - (12{,}300)^2\right)\end{array}}}$$

$$\approx \boxed{1.00}$$

11. Plotting the data shows that the relationship is nonlinear.

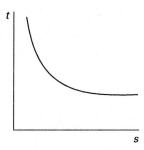

This appears to be an exponential with the form

$$t = ae^{bs}$$

Take the natural log of both sides.

$$\ln t = \ln(ae^{bs})$$
$$= \ln a + \ln(e^{bs})$$
$$= \ln a + bs$$

But, $\ln a$ is just a constant, c.

$$\ln t = c + bs$$

Make the transformation $R = \ln t$.

$$R = c + bs$$

s	R
20	3.76
18	4.95
16	5.95
14	7.00

This is linear.

$$n = 4$$

$$\sum s_i = 20 + 18 + 16 + 14 = 68$$

$$\bar{s} = \frac{\sum s}{n} = \frac{68}{4} = 17$$

$$\sum s_i^2 = (20)^2 + (18)^2 + (16)^2 + (14)^2 = 1176$$

$$\left(\sum s_i\right)^2 = (68)^2 = 4624$$

$$\sum R_i = 3.76 + 4.95 + 5.95 + 7.00 = 21.66$$

$$\bar{R} = \frac{\sum R_i}{n} = \frac{21.66}{4} = 5.415$$

$$\sum R_i^2 = (3.76)^2 + (4.95)^2 + (5.95)^2 + (7.00)^2$$
$$= 123.04$$

$$\left(\sum R_i\right)^2 = (21.66)^2 = 469.16$$

$$\sum s_i R_i = (20)(3.76) + (18)(4.95) + (16)(5.95)$$
$$+ (14)(7.00)$$
$$= 357.5$$

The slope, b, of the transformed line is

$$b = \frac{n\sum s_i R_i - \sum s_i \sum R_i}{n\sum s_i^2 - \left(\sum s_i\right)^2}$$

$$= \frac{(4)(357.5) - (68)(21.66)}{(4)(1176) - (68)^2}$$

$$= -0.536$$

The intercept is

$$c = \bar{R} - b\bar{s}$$
$$= 5.415 - (-0.536)(17)$$
$$= 14.527$$

The transformed equation is

$$R = c + bs$$
$$= 14.527 - 0.536s$$

$$\boxed{\ln t = 14.527 - 0.536s}$$

12. The first step is to graph the data.

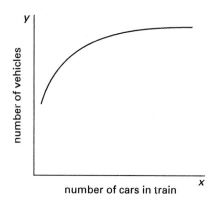

It is assumed that the relationship between the variables has the form $y = a + b\log x$. Therefore, the variable change $z = \log x$ is made, resulting in the following set of data.

z	y
0.301	14.8
0.699	18.0
0.903	20.4
1.079	23.0
1.431	29.9

$$\sum z_i = 4.413$$

$$\sum y_i = 106.1$$

$$\sum z_i^2 = 4.6082$$

$$\sum y_i^2 = 2382.2$$

$$\left(\sum z_i\right)^2 = 19.475$$

$$\left(\sum y_i\right)^2 = 11{,}257.2$$

$$\bar{z} = 0.8826$$

$$\bar{y} = 21.22$$

$$\sum z_i y_i = 103.06$$

$$n = 5$$

The slope is

$$m = \frac{n\sum z_i y_i - \sum z_i \sum y_i}{n\sum z_i^2 - \left(\sum z_i\right)^2}$$

$$= \frac{(5)(103.06) - (4.413)(106.1)}{(5)(4.6082) - 19.475}$$

$$= 13.20$$

The y-intercept is

$$b = \bar{y} - m\bar{z}$$
$$= 21.22 - (13.20)(0.8826)$$
$$= 9.570$$

The resulting equation is

$$y = 9.570 + 13.20z$$

The relationship between x and y is approximately

$$\boxed{y = 9.570 + 13.20 \log x}$$

This is not an optimal correlation, as better correlation coefficients can be obtained if other assumptions about the form of the equation are made. For example, $y = 9.1 + 4\sqrt{x}$ has a better correlation coefficient.

13. (a) Plot the data to verify that they are linear.

x $T - 200$	y $Y - 90$
7.1	2.30
10.3	2.58
0.4	1.56
1.1	1.63
3.4	1.83

step 1:

$$\sum x_i = 22.3 \qquad \sum y_i = 9.9$$
$$\sum x_i^2 = 169.43 \qquad \sum y_i^2 = 20.39$$
$$(\sum x_i)^2 = 497.29 \qquad (\sum y_i)^2 = 98.01$$
$$\bar{x} = \frac{22.3}{5} = 4.46 \qquad \bar{y} = 1.98$$
$$\sum x_i y_i = 51.54$$

step 2: From Eq. 12.70, the slope is

$$m = \frac{n \sum (x_i y_i) - (\sum x_i)(\sum y_i)}{n \sum x_c^2 - (\sum x_i)^2}$$
$$= \frac{(5)(51.54) - (22.3)(9.9)}{(5)(169.43) - 497.29} = 0.1055$$

step 3: From Eq. 12.71, the y-intercept is

$$b = \bar{y} - m\bar{x}$$
$$= 1.98 - (0.1055)(4.46) = 1.509$$

The equation of the line is

$$y = 0.1055x + 1.509$$
$$Y - 90 = 0.1055(T - 200) + 1.509$$
$$Y = \boxed{0.1055T + 70.409}$$

(b) *step 4:* Use Eq. 12.72 to get the correlation coefficient.

$$r = \frac{n \sum (x_i y_i) - (\sum x_i)(\sum y_i)}{\sqrt{\left(n \sum x_i^2 - (\sum x_i)^2\right)\left(n \sum y_i^2 - (\sum y_i)^2\right)}}$$
$$= \frac{(5)(51.54) - (22.3)(9.9)}{\sqrt{\left((5)(169.43) - 497.29\right)\left((5)(20.39) - 98.01\right)}}$$
$$= \boxed{0.995}$$

14. Plot the data to see if they are linear.

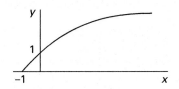

This looks like it could be of the form

$$y = a + b\sqrt{x}$$

However, when x is negative (as in the first point), the function is imaginary. Try shifting the curve to the right, replacing x with $x + 1$.

$$y = a + bz$$
$$z = \sqrt{x + 1}$$

z	y
0	0
1	1
1.414	1.4
1.732	1.7
2	2
2.236	2.2
2.45	2.4
2.65	2.6
2.83	2.8
3	3

Because $y \approx z$, the relationship is

$$\boxed{y = \sqrt{x + 1}}$$

In this problem, the answer was found with some good guesses. Usually, regression would be necessary.

15. (a)

R (Ω)	f	fR (Ω)	fR^2 (Ω^2)
0.200	1	0.200	0.0400
0.210	3	0.630	0.1323
0.220	5	1.100	0.2420
0.230	10	2.300	0.5290
0.240	17	4.080	0.9792
0.250	40	10.000	2.5000
0.260	13	3.380	0.8788
0.270	6	1.620	0.4374
0.280	3	0.840	0.2352
0.290	2	0.580	0.1682
	100	24.730	6.1421

$$\overline{R} = \frac{\sum fR}{\sum f} = \frac{24.730 \ \Omega}{100} = \boxed{0.2473 \ \Omega \quad (0.247 \ \Omega)}$$

The answer is (B).

(b) The sample standard deviation is given by Eq. 12.61.

$$s = \sqrt{\frac{\sum fR^2 - \dfrac{(\sum fR)^2}{n}}{n-1}}$$

$$= \sqrt{\frac{6.1421 \ \Omega - \dfrac{(24.73 \ \Omega)^2}{100}}{99}}$$

$$= \boxed{0.0163 \ \Omega \quad (0.016 \ \Omega)}$$

The answer is (C).

(c) 50% of the observations are below the median resistance, R_{median}. 36 observations are below 0.240. 14 more are needed.

$$R_{\text{median}} = R_{36} + (f_{\text{interval}})(\Delta R_{\text{interval}})$$

$$= 0.240 \ \Omega + \left(\frac{14}{40}\right)(0.250 \ \Omega - 0.240 \ \Omega)$$

$$= \boxed{0.2435 \ \Omega \quad (0.244 \ \Omega)}$$

The answer is (B).

(d) $s^2 = (0.0163 \ \Omega)^2$

$$= \boxed{0.0002656 \ \Omega^2 \quad (0.00027 \ \Omega^2)}$$

The answer is (A).

13 Computer Mathematics

PRACTICE PROBLEMS

1. What is the base-10 value of the base-2 number 1010?

    (A) 3_{10}

    (B) 7_{10}

    (C) 10_{10}

    (D) 20_{10}

2. What is the value of 144_{10} in base-2?

    (A) $1\,001\,000_2$

    (B) $1\,010\,000_2$

    (C) $10\,010\,000_2$

    (D) $10\,010\,001_2$

3. What is the value of 144_{10} in base-8, or octal, numbering?

    (A) 22_8

    (B) 220_8

    (C) 202_8

    (D) 221_8

4. What is the base-10 value of the hexadecimal number $3C1_{16}$?

    (A) 192_{10}

    (B) 768_{10}

    (C) 961_{10}

    (D) $15\,360_{10}$

5. What is the one's complement value of $101\,011_2$?

    (A) $10\,010_2$

    (B) $10\,100_2$

    (C) $10\,111_2$

    (D) $101\,011_2$

SOLUTIONS

1. For a positional numbering system

$$(a_n a_{n-1} \cdots a_2 a_1 a_0) = a_n b^n + a_{n-1} b_{n-1} + \cdots$$
$$+ a_2 b^2 + a_1 b + a_0$$

Therefore, for $b = 2$

$$(1010)_2 = (1)(2^3) + (0)(2^2) + (1)(2) + 0$$
$$= 8 + 0 + 2 + 0$$
$$= \boxed{10_{10}}$$

The answer is (C).

2. Using the remainder method

$144/2 = 72$ remainder 0
$72/2 = 36$ remainder 0
$36/2 = 18$ remainder 0
$18/2 = 9$ remainder 0
$9/2 = 4$ remainder 1
$4/2 = 2$ remainder 0
$2/2 = 1$ remainder 0
$1/2 = 0$ remainder 1

> Reading up the remainder values from the last division remainder of 1 to the first remainder gives a base-2 value of $10\,010\,000_2$. Thus, the binary representation of 144_{10} is $10\,010\,000_2$.

The answer is (C).

3. Using the remainder method gives

$144/8 = 18$ remainder 0
$18/8 = 2$ remainder 2
$2/8 = 0$ remainder 2

> Reading up the remainder values from the last division remainder of 1 to the first remainder gives a base-8 value of 220_8. The octal representation of 144_{10} is 220_8.

The answer is (B).

4. For a position numbering system

$$(a_n a_{n-1} \cdots a_2 a_1 a_0) = a_n b^n + a_{n-1} b_{n-1} + \cdots$$
$$+ a_2 b^2 + a_1 b + a_0$$

Therefore, for $b = 16$,

$$(3C1)_{16} = (3)(16^2) + (12)(16) + 1$$
$$= 768 + 192 + 1$$
$$= \boxed{961_{10}}$$

The base-10 representation of $3C1_{16}$ is 961_{10}.

The answer is (C).

5. The one's complement is

$$N_1^x = N_2^x - 1$$

The one's complement is found by subtracting 1 from the two's complement value. Or, the one's complement can be found by reversing all the digits in the base-2 number. That is, all 0's become 1's and vice versa. Thus, for $101\,011_2$, the one's complement is $010\,100_2$ or, dropping the leading 0, $10\,100_2$.

The answer is (B).

14 Numerical Analysis

PRACTICE PROBLEMS

1. A function is given as $y = 3x^{0.93} + 4.2$. What is the percent error if the value of y at $x = 2.7$ is found by using straight-line interpolation between $x = 2$ and $x = 3$?

(A) 0.060%
(B) 0.18%
(C) 2.5%
(D) 5.4%

2. Given the following data points, find y by straight-line interpolation for $x = 2.75$.

x	y
1	4
2	6
3	2
4	−14

(A) 2.1
(B) 2.4
(C) 2.7
(D) 3.0

3. Using the bisection method, find all the roots of $f(x) = 0$, to the nearest 0.000005.

$$f(x) = x^3 + 2x^2 + 8x - 2$$

4. What is most nearly the root of the following equation?

$$f(x) = x^3 - 4x - 5$$

Use Newton's method with $x_0 = 2$.

(A) 2.47
(B) 2.60
(C) 2.63
(D) 16.7

5. What is the primary disadvantage of performing a straight-line interpolation—as opposed to performing a polynomial interpolation using either the method of Lagrange or Newton?

(A) complicated calculations
(B) ignores curvature
(C) requires iterations
(D) slow process

SOLUTIONS

1. The actual value at $x = 2.7$ is

$$y(x) = 3x^{0.93} + 4.2$$
$$y(2.7) = (3)(2.7)^{0.93} + 4.2$$
$$= 11.756$$

At $x = 3$,
$$y(3) = (3)(3)^{0.93} + 4.2$$
$$= 12.534$$

At $x = 2$,
$$y(2) = (3)(2)^{0.93} + 4.2$$
$$= 9.916$$

Use straight-line interpolation.

$$\frac{x_2 - x}{x_2 - x_1} = \frac{y_2 - y}{y_2 - y_1}$$
$$\frac{3 - 2.7}{3 - 2} = \frac{12.534 - y}{12.534 - 9.916}$$
$$y = 11.749$$

The relative error is given by

$$\frac{\text{actual value} - \text{predicted value}}{\text{actual value}} = \frac{11.756 - 11.749}{11.756}$$
$$= \boxed{0.000595 \ (0.060\%)}$$

The answer is (A).

2. Let $x_1 = 2$; therefore, from the table of data points, $y_1 = 6$. Let $x_2 = 3$; therefore, from the table of data points, $y_2 = 2$.

Let $x = 2.75$. By straight-line interpolation,

$$\frac{x_2 - x}{x_2 - x_1} = \frac{y_2 - y}{y_2 - y_1}$$
$$\frac{3 - 2.75}{3 - 2} = \frac{2 - y}{2 - 6}$$
$$\boxed{y = 3}$$

The answer is (D).

3. $f(x) = x^3 + 2x^2 + 8x - 2$

Try to find an interval in which there is a root.

x	$f(x)$
0	-2
1	9

A root exists in the interval $[0,1]$.

Try $x = \left(\frac{1}{2}\right)(0 + 1) = 0.5$.
$$f(0.5) = (0.5)^3 + (2)(0.5)^2 + (8)(0.5) - 2 = 2.625$$

A root exists in $[0,0.5]$.

Try $x = 0.25$.
$$f(0.25) = (0.25)^3 + (2)(0.25)^2 + (8)(0.25) - 2 = 0.1406$$

A root exists in $[0,0.25]$.

Try $x = 0.125$.
$$f(0.125) = (0.125)^3 + (2)(0.125)^2 + (8)(0.125) - 2$$
$$= -0.967$$

A root exists in $[0.125,0.25]$.

Try $x = \left(\frac{1}{2}\right)(0.125 + 0.25) = 0.1875$.
Continuing,
$$f(0.1875) = -0.42 \quad [0.1875,0.25]$$
$$f(0.21875) = -0.144 \quad [0.21875,0.25]$$
$$f(0.234375) = -0.002 \quad [\text{This is close enough.}]$$

One root is $x_1 \approx \boxed{0.234375}$.

Try to find the other two roots. Use long division to factor the polynomial.

$$
\begin{array}{r}
x^2 + 2.234375x + 8.52368 \\
x - 0.234375 \overline{)\ x^3 + 2x^2 + 8x - 2} \\
-(x^3 - 0.234375x^2) \\
\overline{2.234375x^2 + 8x} \\
-(2.234375x^2 - 0.52368x) \\
\overline{8.52368x - 2} \\
-(8.52368x - 1.9977) \\
\overline{\approx 0}
\end{array}
$$

Use the quadratic equation to find the roots of $x^2 + 2.234375x + 8.52368$.

$$x_2, x_3 = \frac{-2.234375 \pm \sqrt{(2.234375)^2 - (4)(1)(8.52368)}}{(2)(1)}$$
$$= \boxed{-1.117189 \pm j2.697327} \quad [\text{both imaginary}]$$

4. Newton's method for finding a root is

$$x_{n+1} = g(x_n) = x_n - \frac{f(x_n)}{f'(x_n)}$$

The function and its first derivative are

$$f(x) = x^3 - 4x - 5$$
$$f'(x) = 3x^2 - 4$$

The first iteration, with $n = 0$, gives

$$x_0 = 2$$
$$f(x_0) = f(2) = 2^3 - (4)(2) - 5 = -5$$
$$f'(x_0) = f'(2) = (3)(2)^2 - 4 = 8$$
$$x_1 = x_0 - \frac{f(x_0)}{f'(x_0)}$$
$$= 2 - \left(\frac{-5}{8}\right)$$
$$= 2.625$$

The second iteration, with $n = 1$, gives

$$x_1 = 2.625$$
$$f(x_1) = f(2.625) = 2.625^3 - (4)(2.625) - 5 = 2.588$$
$$f'(x_1) = f'(2.625) = (3)(2.625)^2 - 4 = 16.672$$
$$x_2 = x_1 - \frac{f(x_1)}{f'(x_1)}$$
$$= 2.625 - \left(\frac{2.588}{16.672}\right)$$
$$= \boxed{2.470 \quad (2.47)}$$

The iterations may be continued to the desired accuracy.

The answer is (A).

5. The straight-line interpolation technique ignores the effects of curvature. Lagrangian and Newtonian polynomial methods of interpolation do not.

The answer is (B).

15 Advanced Engineering Mathematics

PRACTICE PROBLEMS

1. What is the correct description of the following differential equation?

$$2y''(t) + 3y'(t) + 1 = f(t)$$

(A) linear, first order, homogeneous
(B) nonlinear, first order, homogeneous
(C) linear, second order, nonhomogeneous
(D) nonlinear, second order, nonhomogeneous

2. What is the characteristic equation for the following differential equation?

$$3y''(t) + 5y'(t) + 6y(t) = 24t$$

(A) $3\lambda^2 + 5\lambda + 6 = 0$
(B) $3\lambda^2 + 5\lambda + 6 = 24$
(C) $3\lambda^2 + 5\lambda = 18$
(D) $3\lambda^2 + 5\lambda = 24$

3. What type of function is shown?

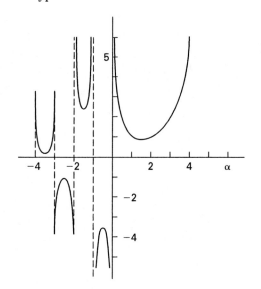

(A) delta function
(B) error function
(C) gamma function
(D) unit finite impulse function

4. Which of the following is primarily used to analyze periodic signals when $s = j\omega$?

(A) Fourier series
(B) Fourier transforms
(C) Laplace transforms
(D) z-transforms

5. What is the Laplace transform of the following?

$$x(t - t_0)$$

(A) $X(s)$
(B) $e^{-st}X(s)$
(C) e^{-st_0}
(D) $e^{-st_0}X(s)$

SOLUTIONS

1. All the coefficients of the terms are constants, therefore, the equation is linear. Because there is a second derivative, the equation is of second order. Because the forcing function, $f(t)$, exists, the implication is that the value of $f(t)$ is non-zero. Therefore, the equation is nonhomogeneous.

The answer is (C).

2. The characteristic equation is

$$3\lambda^2 + 5\lambda + 6 = 0$$

The characteristic equation is used to solve the homogeneous differential equation.

The answer is (A).

3. The illustration shows the factorial function, more commonly called the gamma function.

$$\Gamma(\alpha) = \int_0^\infty e^{-t}t^{(\alpha-1)}dt \quad [\alpha > 0]$$

The answer is (C).

4. Fourier series and Laplace transforms are used to analyze periodic signals. The Laplace transform is more general in nature.

$$\mathcal{L}\big(f(t)\big)\big|_{s=jw} = F(s)\big|_{s=jw} = \mathcal{F}\big(f(t)\big)$$

Therefore, the Fourier series is used.

The Fourier series is equivalent to the Laplace transform when $s = j\omega$, that is, when a transient condition is not present. When $s = \sigma + j\omega$, a transient condition is present, represented by the σ term, and the Fourier series is not equivalent to the Laplace transform.

The answer is (A).

5. The Laplace transform of a time-shifted function is

$$e^{st_0}X(s)$$

The answer is (D).

16 Electromagnetic Theory

PRACTICE PROBLEMS

1. What is the magnitude of the electric force between a proton and electron 0.046 nm apart in free space?
- (A) 9.651×10^{-19} N
- (B) -5.018×10^{-18} N
- (C) 5.018×10^{-18} N
- (D) 1.090×10^{-7} N

2. What is the magnitude of the electric field strength for a proton in free space at 0.046 nm?
- (A) 3.047×10^{-30} N/C
- (B) 1.440×10^{-9} N/C
- (C) 6.805×10^{11} N/C
- (D) 31.30 N/C

3. What is the magnitude of the electric field intensity for a parallel plate capacitor with a potential of 12 V applied and a distance of 0.3 cm between plates?
- (A) 0.12 V/m
- (B) 0.40 V/m
- (C) 40 V/m
- (D) 4000 V/m

4. During design discussion, quartz is picked as the dielectric for certain capacitors. What is the expected permittivity?
- (A) 1.8×10^{-12} C^2/N·m^2
- (B) 8.8×10^{-12} C^2/N·m^2
- (C) 4.4×10^{-11} C^2/N·m^2
- (D) 8.8×10^{-8} C^2/N·m^2

5. What is the magnitude of the electric flux density 0.5 m from a wire with a uniform line charge of 30×10^{-5} C/m?
- (A) 9.5×10^{-5} C/m^2
- (B) 1.9×10^{-4} C/m^2
- (C) 6.0×10^{-4} C/m^2
- (D) 1.7×10^{-3} C/m^2

SOLUTIONS

1. The magnitude of the electric force is given by Coulomb's law. From Eq. 16.3,

$$F = \frac{|Q||Q|}{4\pi\epsilon_0 r^2}$$

$$= \frac{\left(1.6022 \times 10^{-19} \text{ C}\right)\left(1.6022 \times 10^{-19} \text{ C}\right)}{4\pi \left(8.854 \times 10^{-12} \dfrac{\text{C}^2}{\text{N·m}^2}\right)(0.046 \times 10^{-9} \text{ m})^2}$$

$$= \boxed{1.090 \times 10^{-7} \text{ N}}$$

The answer is (D).

Note that the distance 0.046 nm is the approximate distance between the proton and electron in the hydrogen atom.

2. From Eq. 16.6, the magnitude of the electric field strength, E, is

$$E = \frac{Q}{4\pi\epsilon_0 r^2}$$

$$= \frac{1.6022 \times 10^{-19} \text{ C}}{4\pi \left(8.854 \times 10^{-12} \dfrac{\text{C}^2}{\text{N·m}^2}\right)(0.046 \times 10^{-9} \text{ m})^2}$$

$$= \boxed{6.805 \times 10^{11} \text{ N/C}}$$

The answer is (C).

3. From Table 16.6, the electric field intensity for a parallel plate capacitor (ignoring the effects of fringing) is

$$E = \frac{V}{r}$$

$$= \frac{12 \text{ V}}{(0.3 \text{ cm})\left(\dfrac{1 \text{ m}}{100 \text{ cm}}\right)}$$

$$= \boxed{4000 \text{ V/m}}$$

The answer is (D).

4. From Table 16.7,

$$\epsilon_r = 5$$

From Eq. 16.11, the permittivity is

$$\epsilon = \epsilon_r \epsilon_0$$

$$= (5)\left(8.854 \times 10^{-12}\ \frac{C^2}{N{\cdot}m^2}\right)$$

$$= \boxed{4.4 \times 10^{-11}\ C^2/N{\cdot}m^2}$$

The answer is (C).

5. From Eq. 16.21, the electric flux density for a uniform line charge is

$$D = \frac{\rho_l}{2\pi r}$$

$$= \frac{30 \times 10^{-5}\ \dfrac{C}{m}}{2\pi(0.5\ m)}$$

$$= \boxed{9.5 \times 10^{-5}\ C/m^2}$$

The answer is (A).

Note that 30×10^{-5} C/m correlates approximately with 30 A using an electron speed of 10^5 m/s, which is the thermal limit for electrons in silicon at room temperature.

17 Electronic Theory

PRACTICE PROBLEMS

1. Assuming no kinetic energy will be imparted to the electron, what frequency of incoming radiation is required to free an electron from the surface of silicon that has a work function of 4.8 V?

(A) 5.6×10^4 Hz
(B) 1.2×10^{15} Hz
(C) 3.5×10^{23} Hz
(D) 7.2×10^{33} Hz

2. Using the properties of germanium at room temperature, what is the conductivity of the holes?

(A) 7.6×10^{-7} S/m
(B) 7.6×10^{-2} S/m
(C) 4.8×10^{14} S/m
(D) 9.5×10^{14} S/m

3. Silicon and germanium are widely used in semiconductors. How many valence electrons exist for a given atom of either?

(A) 2
(B) 4
(C) 6
(D) 10

4. In an n-type silicon crystal, the number of electrons is

(A) equal to the number of holes
(B) greater than the number of holes
(C) less than the number of holes
(D) equal to the square of the number of holes

5. The n_i of gallium arsenide is 1.1×10^7 cm^{-3}. If the hole concentration is 1.5×10^2 cm^{-3}, what is the electron concentration?

(A) 1.5×10^2 cm^{-3}
(B) 7.3×10^4 cm^{-3}
(C) 1.1×10^7 cm^{-3}
(D) 8.1×10^{11} cm^{-3}

SOLUTIONS

1. From Eq. 17.6,

$$h\nu = \varphi + \tfrac{1}{2}mv^2$$

From App. 17.A, the work function for silicon is 4.8 V. Substituting into the equation gives

$$\frac{\varphi}{h} = \frac{(4.8 \text{ V})\left(1\frac{\text{J}}{\text{C}\cdot\text{V}}\right)\left(1.6022 \times 10^{-19}\frac{\text{C}}{\text{electron}}\right)}{6.6256 \times 10^{-34} \text{ J}\cdot\text{s}}$$

$$= \boxed{1.2 \times 10^{15} \text{ cycles/s (Hz)}}$$

The answer is (B).

2. From Eq. 17.8,

$$\sigma_h = \rho_h \mu_h$$

From App. 17.A,

$$n_i = 2.5 \times 10^{12} \text{ cm}^{-3} = n_{\text{holes}}$$

$$q_h = 1.6022 \times 10^{-19} \text{ C}$$

Because in an intrinsic semiconductor there are equal numbers of electrons and holes,

$$\rho_h = n_i q_h$$
$$= (2.5 \times 10^{12} \text{ cm}^{-3})(1.6022 \times 10^{-19} \text{ C})$$
$$= 4.0 \times 10^{-7} \text{ C/cm}^3$$

From App. 17.A,

$$\mu_h = 1900 \text{ cm}^2/\text{V}\cdot\text{s}$$

$$\sigma = \rho_h \mu_h$$

$$= \left(4.0 \times 10^{-7}\frac{\text{C}}{\text{cm}^3}\right)\left(1900\frac{\text{cm}^2}{\text{V}\cdot\text{s}}\right)\left(100\frac{\text{cm}}{\text{m}}\right)$$

$$= \boxed{7.6 \times 10^{-2} \text{ S/m}}$$

The answer is (B).

3. Silicon has an electronic structure of

$$1s^2 2s^2 2p^6 3s^2 3p^2$$

Germanium has an electronic structure of

$$1s^2 2s^2 2p^6 3s^2 3p^6 3d^{10} 4s^2 4p^2$$

Both have four valence electrons in the outer shell.

The answer is (B).

4. An n-type crystal is doped with donor impurities. This increases the charge carrier density of electrons, and thereby reduces the number of holes, because by Eq. 17.13,

$$np = n_i^2 = \text{constant}$$

The answer is (B).

5. Using the law of mass action, Eq. 17.13,

$$np = n_i^2$$

$$n = \frac{n_i^2}{p} = \frac{\left(1.1 \times 10^7 \text{ cm}^{-3}\right)^2}{1.5 \times 10^2 \text{ cm}^{-3}}$$

$$= \boxed{8.1 \times 10^{11} \text{ cm}^{-3}}$$

The answer is (D).

18 Communication Theory

PRACTICE PROBLEMS

1. A 60 Hz signal is to be transformed into a digital signal. Determine the ideal sampling rate.

2. Determine the Nyquist interval for a 400 Hz signal.

3. A certain signal has an equivalent rectangular bandwidth of 6 MHz. Determine the response time of the system.

4. A 32-bit computer system has equal probabilities for each of its information elements, x_i, or bits. Determine this probability.

5. Describe entropy as used in information theory.

6. The signal power is doubled at a given transmitter. What is the decibel increase?
- (A) 0.3 dB
- (B) 2 dB
- (C) 3 dB
- (D) 20 dB

7. A certain communication signal uses a bandwidth, BW, of 2 kHz at 101 MHz. What is the classification of this signal?
- (A) narrowband, VLF
- (B) voice-grade, HF
- (C) voice-grade, VHF
- (D) wideband, UHF

8. A certain communication system uses the left-most bit as an even parity bit. Which of the following messages is in error?
- (A) 11101
- (B) 00101
- (C) 10001
- (D) 01011

9. What is the maximum thermal noise voltage generated in a 5 kΩ resistor operating in a 6 MHz bandwidth and designed for a temperature service range of $-10°$C to $60°$C?
- (A) 9.71×10^{-6} V
- (B) 1.14×10^{-5} V
- (C) 2.35×10^{-5} V
- (D) 2.46×10^{-5} V

10. What is the shot noise current for a microphone with a 16 kHz bandwidth drawing 500 mA?
- (A) 2.60×10^{-12} A
- (B) 2.56×10^{-8} A
- (C) 2.53×10^{-8} A
- (D) 5.06×10^{-8} A

SOLUTIONS

1. The ideal sampling rate for a signal of frequency f_I is given by the Nyquist rate, f_S.

From Eq. 18.1,

$$f_S = 2f_I$$
$$= (2)(60 \text{ Hz})$$
$$= \boxed{120 \text{ Hz}}$$

Note that, in practice, the rate would have to be greater than this to avoid a zero output, which would occur if the initial sampling occurred at a value of zero for the analog signal.

2. From Eq. 18.2, the Nyquist interval, T_S, is

$$T_S = \frac{1}{2f_I}$$

The information signal, or message signal, frequency is given as 400 Hz. Therefore,

$$T_S = \frac{1}{(2)(400 \text{ Hz}) \left(\frac{\text{s}^{-1}}{1 \text{ Hz}} \right)}$$
$$= \boxed{0.00125 \text{ s} \quad (1.25 \text{ ms})}$$

3. From Eq. 18.4, the bandwidth, $\Delta\Omega$, and duration or response time, ΔT, are related by the following uncertainty relationship.

$$\Delta T \Delta \Omega = 2\pi$$

Rearranging and substituting,

$$\Delta T = \frac{2\pi}{\Delta\Omega} = \frac{2\pi}{6 \times 10^6 \text{ s}^{-1}}$$
$$= \boxed{1.047 \times 10^{-6} \text{ s}}$$

4. The information content of a message is given as 32 bits. The information content is related to the probability as follows.

From Eq. 18.6,

$$I(x_i) = \log_2 \frac{1}{p(x_i)}$$

Rearranging and substituting,

$$p(x_i) = \frac{1}{\text{antilog}_2 I(x_i)}$$
$$= \frac{1}{\text{antilog}_2 (32)}$$
$$= \boxed{\frac{1}{4.295 \times 10^9}}$$

5. Entropy is the mean or average value of a message. It can be considered a measure of the disorder of a system, or of uncertainty in a message. For a given message of alphabet size A, the entropy is determined by Eq. 18.7.

$$H(X) = \sum_{i=1}^{A} p(x_i) I(x_i)$$

Here $p(x_i)$ is the probability of the piece of information x_i. Entropy is normally given in base 2, in units of bits. In base e, the unit is the nat.

6. The decibel is defined by the following ratio of power, P_{dB}.

$$P_{\text{dB}} = 10 \log_{10} \frac{P_2}{P_1}$$

Here P_1 is a reference power and P_2 is the signal power. Because power doubled, P_2/P_1, equals 2.

$$P_{\text{dB}} = 10 \log_{10} 2$$
$$= \boxed{3 \text{ dB}}$$

The answer is (C).

7. Classification can mean a number of things. Given the possible choices, this signal is a voice-grade channel (BW 300 Hz to 4 kHz) in the VHF range (30 MHz to 300 MHz).

The answer is (C).

8. The message is contained in the bits to the right of the first bit in the signal, the parity bit. The parity represents an even or odd total of ones. The following table is helpful.

	message	parity	signal
(A)	1101	1	11101
(B)	0101	0	00101
(C)	0001	1	10001
(D)	1011	1	11011

By comparison, message (D) is in error.

The answer is (D).

9. From Eq. 18.10, the mean squared thermal noise voltage is

$$\overline{v^2} = 4kTR\,(\text{BW})$$

$$v = \sqrt{4kTR\,(\text{BW})}$$

Here, k is Boltzmann's constant, T is in Kelvins, R is resistance, and BW is the bandwidth. The maximum voltage occurs at the maximum temperature (60°C or 333K), thus

$$v = \sqrt{\begin{array}{l}(4)\left(1.38 \times 10^{-23}\,\dfrac{\text{J}}{\text{K}}\right)(333\text{K}) \\[2mm] \times\,(5 \times 10^3\,\Omega)(6 \times 10^6\,\text{Hz})\end{array}}$$

$$= \boxed{2.35 \times 10^{-5}\,\text{V}}$$

The answer is (C).

10. From Eq. 18.13, the mean squared shot noise current is given in terms of the change per carrier, q.

$$\overline{i_n^2} = 2qI\,(\text{BW})$$

$$i_n = \sqrt{2qI\,(\text{BW})}$$

$$= \sqrt{\begin{array}{l}(2)(1.6022 \times 10^{-19}\,\text{C}) \\[2mm] (500 \times 10^{-3}\,\text{A})(16 \times 10^3\,\text{Hz})\end{array}}$$

$$= \boxed{5.06 \times 10^{-8}\,\text{A}}$$

The answer is (D).

Theory

19 Acoustic and Transducer Theory

PRACTICE PROBLEMS

1. Explain two differences between sound waves and electromagnetic waves.

2. Define the term "transducer."

3. What is the sound pressure level in air for a signal at the pain sensation threshold of 1 W/m²?
- (A) −120 dB
- (B) 3.0 dB
- (C) 60 dB
- (D) 120 dB

4. Explain the difference between photovoltaic and photoconductive transduction.

5. The reference pressure level, p_0, in air of 20 μPa correlates with the threshold audible frequency of 2000 Hz. What is the wavelength of this sound wave?
- (A) 0.17 m
- (B) 1.0 m
- (C) 6.1 m
- (D) 20 m

SOLUTIONS

1. Sound is a longitudinal or compression wave. Electromagnetic waves are transverse. That is, the electric and magnetic fields are at right angles to the direction of propagation. Additionally, sound waves require a medium in which to travel, while electromagnetic waves do not.

2. A transducer is a communication interface that provides a usable output in response to a specified measurand (i.e., a physical quantity, property, or condition).

3. From Ex. 19.2, the pressure level corresponding to 1 W/m² is 29.3 Pa. The reference pressure (corresponding to 0 dB) is defined as 20 μPa. Substituting from Eq. 19.8 gives

$$L_p = 20 \log \frac{p}{p_0}$$

$$= 20 \log \frac{29.3 \,\text{Pa}}{20 \times 10^{-6} \,\text{Pa}}$$

$$= \boxed{123 \text{ dB} \quad (120 \text{ dB})}$$

The answer is (D).

Note that the "pain" threshold is nominally 120 dB in most references.

4. The primary difference between photovoltaic and photoconductive transduction is that a photovoltaic device uses dissimilar metals and a photoconductive device uses a semiconductor pn junction.

5. From Table 19.1, the speed of sound in air is approximately 330 m/s. Substituting this and the given frequency into Eq. 19.1 gives

$$\text{v}_s = f \lambda$$

$$\lambda = \frac{\text{v}_s}{f} = \frac{330 \, \dfrac{\text{m}}{\text{s}}}{2000 \text{ Hz}}$$

$$= \boxed{0.165 \text{ m} \quad (0.17 \text{ m})}$$

The answer is (A).

20 Electrostatics

Field Theory

PRACTICE PROBLEMS

1. Determine the magnitude of the electric field necessary to place a 1 N force on an electron.

2. A point charge of 4.8×10^{-19} C is located at a point $(3,1,0)$. What is the electric field strength at $(3,0,1)$?

(A) 1.5×10^{-9} V/m $\left(\dfrac{-\mathbf{y}}{\sqrt{2}} + \dfrac{\mathbf{z}}{\sqrt{2}}\right)$

(B) 2.2×10^{-9} V/m $\left(\dfrac{-\mathbf{y}}{\sqrt{2}} + \dfrac{\mathbf{z}}{\sqrt{2}}\right)$

(C) 3.1×10^{-9} V/m $(-\mathbf{y} + \mathbf{z})$

(D) 4.3×10^{-9} V/m $(-\mathbf{y} + \mathbf{z})$

3. Water molecules in vapor form have a fractional dipole charge of approximately 5.26×10^{-20} C. If the dipole bond length is 1.17 Å, what is the magnitude of the dipole moment?

(A) 6.2×10^{-32} C·m

(B) 6.2×10^{-30} C·m

(C) 5.3×10^{-20} C·m

(D) 6.1×10^{-20} C·m

4. The potential of a point charge, in spherical coordinates and relative to infinity, is

$$V = \frac{Q}{4\pi\varepsilon_0 r}$$

Determine the expression for the electric field strength, **E**.

5. The electric field in an n-type material outside the space charge region is given by

$$\mathbf{E} = 2\mathbf{x} + 2\mathbf{y} + 3\mathbf{z}$$

Determine the charge density, ρ, and explain the result.

SOLUTIONS

1. The magnitude of the force is

$$F = QE$$

Rearranging gives

$$E = \frac{F}{Q}$$

$$= \frac{1 \text{ N}}{1.602 \times 10^{-19} \text{ C}}$$

$$= 6.2 \times 10^{18} \, \frac{\text{N}}{\text{C}}$$

$$= \boxed{6.2 \times 10^{18} \text{ V/m}}$$

For determining the magnitude of the force and not the direction, the minus sign is not required on the charge.

2.

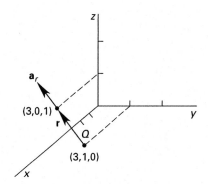

The electric field is

$$\mathbf{E} = \left(\frac{Q}{4\pi\epsilon r^2}\right)\mathbf{a}_r$$

First, determine the value of the unit vector, $\mathbf{a}_r$.

$$a_r = \sqrt{(3-3)^2 + (0-1)^2 + (1-0)^2} = \sqrt{2}$$

$$\mathbf{r} = 0\mathbf{x} - 1\mathbf{y} + 1\mathbf{z}$$

Thus,

$$a_r = \frac{r}{a_r} = \frac{1}{\sqrt{2}}(-\mathbf{y} + \mathbf{z})$$

Substituting this and the given information (free space is assumed) gives

$$\mathbf{E} = \left(\frac{Q}{4\pi\epsilon r^2}\right)\mathbf{a}_r$$

$$= \left(\frac{4.8 \times 10^{-19}\text{ C}}{4\pi\left(8.854 \times 10^{-12}\,\frac{\text{C}^2}{\text{N·m}^2}\right)(\sqrt{2}\text{ m})^2}\right)$$

$$\times \left(\left(\frac{1}{\sqrt{2}}\right)(-\mathbf{y}+\mathbf{z})\right)$$

$$= \boxed{2.16 \times 10^{-9}\text{ N/C}\left(\frac{-\mathbf{y}}{\sqrt{2}} + \frac{\mathbf{z}}{\sqrt{2}}\right)}$$

The answer is (B).

3. The dipole moment vector, given in terms of a single positive charge, is

$$\mathbf{p} = q\mathbf{d}$$

It points from the negative to positive charge. The separation of charges is d.

An angstrom, Å, equals 10^{-10} m. Substituting gives

$$p = qd = (5.26 \times 10^{-20}\text{ C})(1.17 \times 10^{-10}\text{ m})$$

$$= \boxed{6.18 \times 10^{-30}\text{ C·m}\quad (6.2 \times 10^{-30}\text{ C·m})}$$

The answer is (B).

4. The electric field strength is

$$\mathbf{E} = -\nabla V$$

$$= -\nabla\left(\frac{Q}{4\pi\epsilon_0 r}\right)$$

$$= \left(-\frac{Q}{4\pi\epsilon_0}\right)\left(\nabla\frac{1}{r}\right)$$

In spherical coordinates, the del operator can be represented as

$$\nabla = \left(\frac{\partial}{\partial r}\right)\mathbf{r} + \left(\frac{1}{r}\right)\left(\frac{\partial}{\partial\theta}\right)\boldsymbol{\theta} + \left(\frac{1}{r\sin\theta}\right)\left(\frac{\partial}{\partial\phi}\right)\boldsymbol{\phi}$$

For a point charge, the potential and the electric field are constant with respect to θ and ϕ. Thus,

$$\mathbf{E} = \left(-\frac{Q}{4\pi\epsilon_0}\right)\left(\frac{\partial\frac{1}{r}}{\partial r}\right)\mathbf{r}$$

$$= \left(\frac{Q}{4\pi\epsilon_0 r^2}\right)\mathbf{r}$$

$$= \boxed{\left(\frac{Q}{4\pi\epsilon_0 r^2}\right)\mathbf{a}_r}$$

The last equation changes the symbol from $\mathbf{r}$ to $\mathbf{a}_r$ only. Many variations exist.

5. The charge density, ρ, can be determined from Gauss' law.

$$\text{div }\mathbf{E} = \nabla\cdot\mathbf{E} = \frac{\rho}{\epsilon}$$

$$\nabla\cdot\mathbf{E} = \nabla\cdot(2\mathbf{x} + 2\mathbf{y} + 3\mathbf{z})$$

$$= \left(\frac{\partial 2}{\partial x}\right)\mathbf{x} + \left(\frac{\partial 2}{\partial y}\right)\mathbf{y} + \left(\frac{\partial 3}{\partial z}\right)\mathbf{z}$$

$$= 0$$

When the electric field is constant, the divergence is zero. This indicates there is no *net charge* in the region, as is the case for the *pn* junction not under the influence of an external electric field.

21 Electrostatic Fields

PRACTICE PROBLEMS

1. Determine the electric flux density magnitude if the polarization is 2.0×10^{-6} C/m^2 and the susceptibility is 2.2.

    (A) 1.8×10^{-17} C/m^2
    (B) 1.8×10^{-6} C/m^2
    (C) 2.9×10^{-6} C/m^2
    (D) 4.0×10^{-6} C/m^2

2. Given $\mathbf{D} = 2 \times 10^{-6}\ \mathbf{a}_r$ C/m^2 and $\varepsilon_r = 3.0$, what is the electric field strength?

    (A) $6.7 \times 10^{-7}\ \mathbf{a}_r$ N/C
    (B) $2.0 \times 10^{-4}\ \mathbf{a}_r$ N/C
    (C) $7.5 \times 10^{4}\ \mathbf{a}_r$ N/C
    (D) $2.3 \times 10^{5}\ \mathbf{a}_r$ N/C

3. The following values for relative permittivity are taken from a standard reference: paper, 2.0; quartz, 5.0; and marble, 8.3.

The electric field strength is 2.0×10^5 V/m. $\mathbf{D}$ must be greater than or equal to 7×10^{-6} C/m^2. Determine which of the given dielectrics is/are suitable.

4. Moist soil has a conductivity of approximately 10^{-3} S/m and an ε_r of 3.0. A certain electric field strength is given by

$$E = 3.5 \times 10^{-6} \sin\left(2.5 \times 10^3\right) t\ \frac{\text{V}}{\text{m}}$$

Determine the displacement current density.

5. In what circuit element is displacement current largest: conductor, inductor, or capacitor? Explain.

SOLUTIONS

1. The polarization for a material with susceptability χ_e is

$$P = \chi_e \varepsilon_0 E$$

This assumes that the electric field $\mathbf{E}$ and the polarization $\mathbf{P}$ are in the same direction, that is, the material is isotropic and linear. Given that, the electic flux density is

$$D = \varepsilon_0 E + P$$

Substituting P/χ_e for $\varepsilon_0 E$,

$$D = \frac{P}{\chi_e} + P$$

$$= \frac{2 \times 10^{-6}}{2.2}\ \frac{\text{C}}{\text{m}^2} + 2 \times 10^{-6}\ \frac{\text{C}}{\text{m}^2}$$

$$= \boxed{2.9 \times 10^{-6}\ \text{C/m}^2}$$

The answer is (C).

2. The flux density $\mathbf{D}$ in a medium of permittivity ε_r is

$$\mathbf{D} = \varepsilon_0 \varepsilon_r \mathbf{E}$$

Rearranging gives

$$\mathbf{E} = \frac{\mathbf{D}}{\varepsilon_0 \varepsilon_r} = \frac{2 \times 10^{-6}\ \mathbf{a}_r\ \dfrac{\text{C}}{\text{m}^2}}{\left(8.854 \times 10^{-12}\ \dfrac{\text{C}^2}{\text{N·m}^2}\right)(3.0)}$$

$$= \boxed{7.5 \times 10^{4}\ \mathbf{a}_r\ \text{N/C}}$$

The answer is (C).

3. Determine the minimum value of the permittivity ε_r required.

$$\mathbf{D} = \varepsilon_0 \varepsilon_r \mathbf{E}$$

Assuming the dielectrics to be isotropic and linear,

$$D = \varepsilon_0 \varepsilon_r E$$

$$\varepsilon_r = \frac{D}{E \varepsilon_0} = \frac{7 \times 10^{-6}\ \dfrac{\text{C}}{\text{m}^2}}{\left(2.0 \times 10^5\ \dfrac{\text{V}}{\text{m}}\right)\left(8.854 \times 10^{-12}\ \dfrac{\text{C}^2}{\text{N·m}^2}\right)}$$

$$= 3.9$$

Thus, the quartz or marble is an acceptable dielectric.

4. Determine the value of the magnitude of the electric flux density D.

$$D = \varepsilon_0 \varepsilon_r E$$

$$= \left(8.854 \times 10^{-12} \, \frac{C^2}{N \cdot m^2} \right) (3.0)$$

$$\times \left(3.5 \times 10^{-6} \sin (2.5 \times 10^3) \, t \, \frac{V}{m} \right)$$

$$= 9.30 \times 10^{-17} \sin (2.5 \times 10^3) \, t \, C/m^2$$

The displacement current is

$$J_d = \frac{\partial D}{\partial t}$$

$$= \frac{\partial}{\partial t} \left(9.30 \times 10^{-17} \sin (2.5 \times 10^3) \, t \, \frac{C}{m^2} \right)$$

$$= \left(9.30 \times 10^{-17} \right) \left(2.5 \times 10^3 \cos (2.5 \times 10^3) \, t \, \frac{A}{m^2} \right)$$

$$= \boxed{2.32 \times 10^{-13} \cos (2.5 \times 10^3) \, t \, A/m^2}$$

5. The displacement current is largest in a capacitor. The charge stored as a capacitor, of capacitance C, is

$$Q = CV$$

The capacitance is proportional to the area.

$$C \propto A$$

Capacitors, in general, have large areas and store high concentrations of charges. Therefore, the electric field strength is large and, consequently, so is $\mathbf{D}$. At high frequencies this results in a significant displacement current.

22 Magnetostatics

PRACTICE PROBLEMS

1. The flux density a distance r from a current-carrying conductor of length dl varies as

    (A) $\ln r$

    (B) $1/r$

    (C) $1/r^2$

    (D) constant

2. Determine the magnitude of the core flux density for one turn of wire wrapped around a cast-iron core with a relative permeability, μ_r, of 400. The wire is rated at 15 A. The core measures 0.10 m × 0.15 m.

3. In a straight, infinitely long conductor, how much current does it take to produce a magnetic field 1 m away that is equivalent in strength to the Earth's, that is, 5×10^{-5} T?

4. Consider the following illustration.

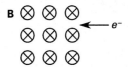

The magnetic field, **B**, is uniform and directed into the paper. An electron is injected into the field from the right with some initial velocity. In what direction does the electron initially deflect?

    (A) up

    (B) down

    (C) left

    (D) right

5. A galvanometer moving-coil, used in instrumentation, is placed between two permanent magnets with a **B**-field strength of 4.0×10^{-3} T in the directions shown. The radius of the coil is 0.5 cm. The coil contains 10 turns, or loops, and has a current rating of 1 mA. Determine the maximum torque that the meter can generate against the restoring spring.

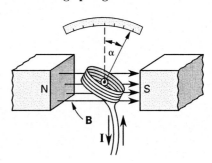

SOLUTIONS

1. The Biot-Savart law gives the magnitude of flux density in differential form as

$$dB = \left(\frac{\mu_0 I dl}{4\pi r^2}\right)\sin\theta$$

The answer is (C).

2. The flux produced by a current I that travels N turns over a distance l is

$$\Phi = \mu N I l = B A$$

Rearranging to solve for the flux density B over area A gives

$$B = \frac{\mu N I l}{A} = \frac{\mu_0 \mu_r N I l}{A}$$

$$= \frac{\left(1.2566 \times 10^{-6}\ \dfrac{\text{N}}{\text{A}^2}\right)(400)(1)(15\ \text{A})}{\times (0.10\ \text{m} + 0.15\ \text{m} + 0.10\ \text{m} + 0.15\ \text{m})}{(0.10\ \text{m})(0.15\ \text{m})}$$

$$= \boxed{0.25\ \text{T}}$$

3. The magnetic field, B, of a straight, infinitely long conductor is

$$B = \frac{\mu I}{2\pi r}$$

Because no specific condition was given, free space permeability, μ_0, is assumed. Rearranging and substituting gives

$$I = \frac{2\pi r B}{\mu_0}$$

$$= \frac{2\pi\,(1\ \text{m})\left(5 \times 10^{-5}\ \text{T}\right)}{1.2566 \times 10^{-6}\ \dfrac{\text{H}}{\text{m}}}$$

$$= \boxed{250\ \text{A}}$$

4. The magnetic force on a moving charged particle q, moving with velocity $\mathbf{v}$ in magnetic field $\mathbf{B}$ is

$$\mathbf{F} = q\mathbf{v} \times \mathbf{B}$$

Using the right-hand rule to determine the direction of the cross-product, the initial deflection appears to be down toward the bottom of the paper. But the electron's charge is negative. Therefore, the deflection is upward toward the top of the paper.

The answer is (A).

5. The torque on magnetic moment $\mathbf{m}$ in field $\mathbf{B}$ is

$$\mathbf{T} = \mathbf{m} \times \mathbf{B}$$

The strength of the $\mathbf{B}$-field is given. The magnetic moment, m, can be determined from

$$m = IA$$

Because there are N turns of wire, each carrying a current I, the equation is

$$m = NIA = NI\pi r^2$$

$$= (10)\,(0.001\ \text{A})\,(\pi)\,(0.005\ \text{m})^2$$

$$= 7.85 \times 10^{-7}\ \text{A·m}^2$$

The maximum magnetic torque can now be determined by substitution.

$$\mathbf{T} = \mathbf{m} \times \mathbf{B}$$

$$T_{\text{max}} = mB\sin\frac{\pi}{2} = mB$$

$$= \left(7.85 \times 10^{-7}\ \text{A·m}^2\right)\left(4.0 \times 10^{-3}\ \text{T}\right)$$

$$= \boxed{3.1 \times 10^{-9}\ \text{N·m}}$$

Magnetostatic Fields

Field Theory

PRACTICE PROBLEMS

1. Define magnetization, and state its mathematical formula with appropriate units.

2. Write the formula for the magnetic flux density that is always valid regardless of the linearity of the material. Use the SI form of the equation.

3. By what factor does the flux density increase if high-purity iron is substituted for commercial iron as the core of a transformer?

    (A) 4
    (B) 30
    (C) 6000
    (D) 2×10^5

4. The magnetic field strength of Earth is approximately 40 A/m. Determine the current required in a single turn of wire located at the equator to generate the same magnetic field at distances far from Earth's surface.

5. What type of magnetic material generates an internal magnetic field that opposes the applied field?

    (A) diamagnetic
    (B) paramagnetic
    (C) ferromagnetic
    (D) antiferromagnetic

SOLUTIONS

1. Magnetization, $\mathbf{M}$, is the magnetic moment per unit volume, and is given by

$$\mathbf{M} = \frac{\mathbf{m}}{v}$$

The units are A/m.

2. The formula gives the magnetic flux density, $\mathbf{B}$, in terms of the magnetic field, $\mathbf{H}$, and the magnetization, $\mathbf{M}$, in the permittivity of free space, μ_0.

$$\boxed{\mathbf{B} = \mu_0 \mathbf{H} + \mu_0 \mathbf{M}}$$

3. The relative permeability, μ_r, of commercial-grade iron is approximately 6000, while that of high-purity iron is 2×10^5. The flux density magnitude is

$$B = \mu_0 \mu_r H$$

The only change from one material to the next is μ_r. Thus, the factor f is

$$f = \frac{\mu_r \text{ (high purity)}}{\mu_r \text{ (commercial)}}$$

$$= \frac{2 \times 10^5}{6000}$$

$$= \boxed{33 \quad (30)}$$

The answer is (B).

4. At distances far from Earth's surface, the applicable equation is

$$H = \frac{NI}{l}$$

Rearranging gives

$$I = \frac{Hl}{N}$$

Earth's radius is approximately 6.4×10^6 m.

The path length is

$$l = 2\pi r = 2\pi (6.4 \times 10^6 \text{ m})$$

$$= 4.0 \times 10^7 \text{ m}$$

Substituting the known and calculated information gives

$$I = \frac{Hl}{N} = \frac{\left(40 \ \dfrac{A}{m}\right)(4.0 \times 10^7 \ m)}{1}$$

$$= \boxed{1.6 \times 10^9 \ A}$$

5. Diamagnetic materials tend to reduce the flux by generating an opposing internal magnetic field.

The answer is (A).

24 Electrodynamics

PRACTICE PROBLEMS

1. The following information is taken from a certain portable generator: v is 1450 m/s, B is 1.2×10^{-2} T, loop length is 0.4 m, and output voltage is 120 V. Estimate the number of turns required to induce the indicated voltage.

(A) 1.0
(B) 17
(C) 18
(D) 120

2. Which law describes induced electromotance?

(A) Faraday's law
(B) Ampere's law
(C) Coulomb's law
(D) Gauss' law

3. Which of the following terms represents the magnetomotive force, or mmf?

(A) $\oint \mathbf{E} \cdot d\mathbf{l}$
(B) $\oint \varepsilon \mathbf{D} \cdot d\mathbf{l}$
(C) $\mu_0 \mathbf{H}$
(D) $\oint \mathbf{H} \cdot d\mathbf{l}$

4. In the following illustration, determine the direction of energy flow.

5. What fundamental action must occur to a charge in order to produce an electromagnetic wave capable of transporting energy in free space?

SOLUTIONS

1. The induced voltage resulting from a conductor moving in a uniform magnetic field is

$$v = -NBl\mathbf{v}$$

Rearranging and solving gives

$$N = \left| \frac{v}{Bl\mathbf{v}} \right|$$

$$= \frac{120 \text{ V}}{\left(1.2 \times 10^{-2} \text{ T}\right)\left(0.4 \text{ m}\right)\left(1450 \, \dfrac{\text{m}}{\text{s}}\right)}$$

$$= 17.2$$

Therefore, 18 turns are required as a minimum.

The answer is (C).

2. Faraday's law of magnetic induction is used to determine the amount of induced voltage, also called induced electromotance.

The answer is (A).

3. The mmf is the integral of the magnetic field strength, $\mathbf{H}$, along the path, $d\mathbf{l}$, where the magnetic flux is present. It is often given as NI, the number of turns times the current. The force produced per unit flux set up in the conductor is $\mathbf{H}$, hence the name.

The answer is (D).

4. The direction of energy flow, or power density, is given by Poynting's vector, $\mathbf{S}$.

$$\mathbf{S} = \mathbf{E} \times \mathbf{H}$$

In terms of the $\mathbf{B}$-field, assuming a linear, isotropic medium,

$$\mathbf{S} = \frac{1}{\mu} \mathbf{E} \times \mathbf{B}$$

If the material were not isotropic, the permeability would remain with the $\mathbf{B}$-field as part of the cross product. The direction of $\mathbf{S}$ would then depend on the product of the permeability and magnetic flux density fields, $\mu\mathbf{B}$, as well.

Using the right-hand rule, the direction of the cross product **S** is into the plane of the paper, ⊗.

5. The charge must be accelerated or decelerated by some means.

Note then that a stationary charge is related to an electric field, a charge in motion with uniform velocity is related to a magnetic field, and an accelerating charge is related to a radiated field.

25 Maxwell's Equations

PRACTICE PROBLEMS

1. Which of the following point forms of Maxwell's equations implies the nonexistence of magnetic monopoles?

(A) $\nabla \cdot \mathbf{D} = \rho$

(B) $\nabla \cdot \mathbf{B} = 0$

(C) $\nabla \times \mathbf{E} = -\dfrac{\partial \mathbf{B}}{\partial t}$

(D) $\nabla \cdot \mathbf{D} = 0$

2. What is missing from the following equation?

$$\nabla \times \mathbf{H} = \mathbf{J}_c + \frac{\partial}{\partial t}$$

(A) **B**

(B) **D**

(C) **E**

(D) **H**

3. Which of the following equations for magnetic circuits is analogous to the equation shown for an electric circuit?

$$\mathrm{emf} = V = IR$$

(A) $G = \dfrac{1}{R}$

(B) $P_m = \dfrac{\mu A}{l}$

(C) $\mathcal{R} = \dfrac{l}{\mu A}$

(D) $V_m = \phi \mathcal{R}$

4. Which of the following is the continuity equation?

(A) $\nabla \cdot \mathbf{D} = \rho$

(B) $\mathbf{J} = \sigma \mathbf{E}$

(C) $\nabla \times \mathbf{J} = -\dfrac{\partial \rho}{\partial t}$

(D) $\mathbf{F} = Q(\mathbf{E} + \mathbf{v} + \mathbf{B})$

5. What is missing in the following equation?

$$\nabla \times \mathbf{E} = -\frac{\partial}{\partial t}$$

(A) **B**

(B) **D**

(C) **H**

(D) **J**

SOLUTIONS

1. The nonexistence of magnetic monopoles is implied by the divergence of the magnetic flux density, **B**, being equal to zero.

$$\nabla \cdot \mathbf{B} = 0$$

The answer is (B).

2.
$$\nabla \times \mathbf{H} = \mathbf{J}_c + \mathbf{J}_d$$
$$\mathbf{J}_d = \frac{\partial \mathbf{D}}{\partial t}$$

The electric flux density, **D**, is missing from the displacement current term.

The answer is (B).

3. The magnetomotive force (mmf) is analogous to the electromotive force (emf). The analogous magnetic equation is

$$\text{mmf} = V_m = \phi \mathcal{R}$$

The answer is (D).

4. The continuity equation states that the divergence of the current density equals the negative rate of change of the charge density.

$$\nabla \times \mathbf{J} = -\frac{\partial \rho}{\partial t}$$

The answer is (C).

5. The curl of the electric field is equal to the negative rate of change of the magnetic flux density.

$$\nabla \times \mathbf{E} = -\frac{\partial \mathbf{B}}{\partial t}$$

Thus, the magnetic flux density term, **B**, is missing.

The answer is (A).

26 DC Circuit Fundamentals

PRACTICE PROBLEMS

1. What is the power dissipated in R_3 of the circuit shown?

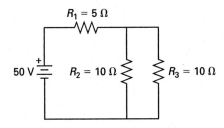

$R_1 = 5\ \Omega$

50 V $R_2 = 10\ \Omega$ $R_3 = 10\ \Omega$

2. Calculate the resistance of the load necessary for maximum power transfer to occur in the circuit shown.

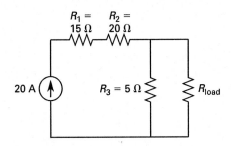

$R_1 = 15\ \Omega$ $R_2 = 20\ \Omega$

20 A $R_3 = 5\ \Omega$ R_{load}

3. For the circuit shown, which method will result in fewer equations: the loop-current method or the node-voltage method?

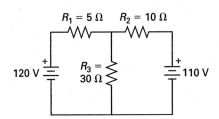

$R_1 = 5\ \Omega$ $R_2 = 10\ \Omega$

120 V $R_3 = 30\ \Omega$ 110 V

4. Using the loop-current method, solve for the current in R_3 in Prob. 3.

5. Using the node-voltage method, solve for the current in R_3 in Prob. 3.

SOLUTIONS

1. Power is given by

$$P = VI = I^2 R = \frac{V^2}{R}$$

Using the last form of the equation, determine the voltage across R_3. Combine the parallel resistors R_2 and R_3.

$$R_{2/3} = \frac{R_2 R_3}{R_2 + R_3} = \frac{(10\ \Omega)(10\ \Omega)}{10\ \Omega + 10\ \Omega} = 5\ \Omega$$

The circuit is transformed as follows.

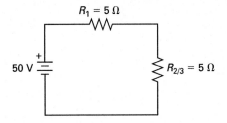

$R_1 = 5\ \Omega$

50 V $R_{2/3} = 5\ \Omega$

Using the voltage-divider concept, the voltage across $R_{2/3}$ (and R_3) is

$$V_{2/3} = V_3 = 50\text{ V} \left(\frac{R_{2/3}}{R_1 + R_{2/3}} \right)$$

$$= (50\text{ V}) \left(\frac{5\ \Omega}{5\ \Omega + 5\ \Omega} \right)$$

$$= 25\text{ V}$$

The power dissipated in R_3 is

$$P_3 = \frac{V_3^2}{R_3} = \frac{(25\text{ V})^2}{10\ \Omega}$$

$$= \boxed{62\text{ W}}$$

2. Maximum power transfer occurs when the load resistance equals the Thevenin or Norton equivalent resistance.

Open circuit the current source and determine R_{Th}.

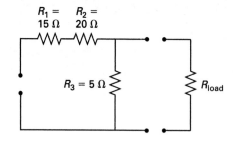

The 15 Ω and 20 Ω resistors are in an open-circuit path. Thus,

$$R_{Th} = R_3 = \boxed{5 \ \Omega}$$

3. The loop-current method using Kirchoff's voltage law (KVL) uses $n-1$ equations, where n is the number of loops. There are three loops present. The number of equations required is two.

The node-voltage method using Kirchoff's current law (KCL) requires $n-1$ equations, where n is the number of principal nodes. There are two principal nodes. Only one equation is required.

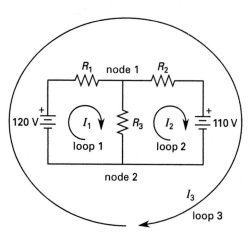

The node-voltage method requires fewer equations.

4. Consider the following circuit.

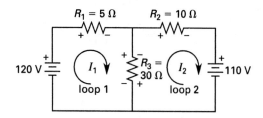

Using the selected loops with polarities for the resistive elements based on the loop-current direction, KVL for loop 1 is

$$120 \text{ V} - I_1 R_1 - (I_1 - I_2) R_3 = 0 \text{ V}$$

Rearranging gives

$$I_1 (R_1 + R_3) - I_2 R_3 = 120 \text{ V} \qquad \text{[I]}$$

KVL for loop 2 is

$$-(I_2 - I_1) R_3 - I_2 R_2 - 110 \text{ V} = 0 \text{ V}$$

Rearranging gives

$$I_1 R_3 - I_2 (R_3 + R_2) = 110 \text{ V} \qquad \text{[II]}$$

There are now two equations, [I] and [II], with two unknowns, I_1 and I_2. Use Cramer's rule and matrix algebra to solve for I_1 and I_2.

$$\begin{vmatrix} R_1 + R_3 & -R_3 \\ R_3 & -(R_3 + R_2) \end{vmatrix} \begin{vmatrix} I_1 \\ I_2 \end{vmatrix} = \begin{vmatrix} 120 \text{ A} \\ 110 \text{ A} \end{vmatrix}$$

$$\begin{vmatrix} 35 \text{ A} & -30 \text{ A} \\ 30 \text{ A} & -40 \text{ A} \end{vmatrix} \begin{vmatrix} I_1 \\ I_2 \end{vmatrix} = \begin{vmatrix} 120 \text{ A} \\ 110 \text{ A} \end{vmatrix}$$

$$I_1 = \frac{\begin{vmatrix} 120 \text{ A} & -30 \text{ A} \\ 110 \text{ A} & -40 \text{ A} \end{vmatrix}}{\begin{vmatrix} 35 \text{ A} & -30 \text{ A} \\ 30 \text{ A} & -40 \text{ A} \end{vmatrix}} = \frac{-1500 \text{ A}}{-500 \text{ A}} = 3 \text{ A}$$

$$I_2 = \frac{\begin{vmatrix} 35 \text{ A} & 120 \text{ A} \\ 30 \text{ A} & 110 \text{ A} \end{vmatrix}}{\begin{vmatrix} 35 \text{ A} & -30 \text{ A} \\ 30 \text{ A} & -40 \text{ A} \end{vmatrix}} = \frac{250 \text{ A}}{-500 \text{ A}} = -0.5 \text{ A}$$

Summing to obtain the current through R_3 and noting that the direction of I_2 was incorrect gives

$$I_3 = |I_1| + |I_2| = 3 \text{ A} + 0.5 \text{ A}$$

$$= \boxed{3.5 \text{ A}}$$

Note that the direction of the current I_3 is identical to the direction of I_1, because it is positive.

5. Consider the following circuit.

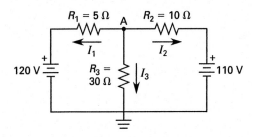

The principal node A is labeled. The node at the lower portion of the circuit is chosen as the reference or ground node. All currents are assumed to be positive.

Writing KCL for node A gives

$$\frac{V_A - 120 \text{ V}}{R_1} + \frac{V_A - 0 \text{ V}}{R_3} + \frac{V_A - 110 \text{ V}}{R_2} = 0 \text{ V}$$

Solve for V_A.

$$\frac{V_A}{R_1} - \frac{120 \text{ V}}{R_1} + \frac{V_A}{R_3} + \frac{V_A}{R_2} - \frac{110 \text{ V}}{R_2} = 0 \text{ V}$$

$$\frac{V_A}{R_1} + \frac{V_A}{R_3} + \frac{V_A}{R_2} = \frac{120 \text{ V}}{R_1} + \frac{110 \text{ V}}{R_2}$$

$$V_A = \frac{\dfrac{120 \text{ V}}{R_1} + \dfrac{110 \text{ V}}{R_2}}{\dfrac{1}{R_1} + \dfrac{1}{R_3} + \dfrac{1}{R_2}}$$

$$= \frac{\dfrac{120 \text{ V}}{5 \text{ }\Omega} + \dfrac{110 \text{ V}}{10 \text{ }\Omega}}{\dfrac{1}{5 \text{ }\Omega} + \dfrac{1}{30 \text{ }\Omega} + \dfrac{1}{10 \text{ }\Omega}}$$

$$= 105 \text{ V}$$

Use V_A to determine the desired current.

$$I_3 = \frac{V_A}{R_3} = \frac{105 \text{ V}}{30 \text{ }\Omega} = \boxed{3.5 \text{ A}}$$

Circuit Theory

27 AC Circuit Fundamentals

PRACTICE PROBLEMS

1. Express the phasor voltage 120 V∠30° as a function of time if the frequency is 60 Hz. The voltage is given in "effective value phasor" notation.

2. Express each of the following in rectangular form.

(a) $15∠-180°$

(b) $10∠37°$

(c) $50∠120°$

(d) $21∠-90°$

3. Express each of the following in phasor form.

(a) $(6 + j7)$

(b) $(50 - j60)$

(c) $(-75 + j45)$

(d) $(90 - j180)$

4. Calculate the following, and express each in phasor form.

(a) $\dfrac{7 + j5}{6}$

(b) $\dfrac{13 + j17}{15 - j10}$

(c) $\dfrac{0.020∠90°}{0.034∠56°}$

5. A sinusoidal waveform has a peak value at $t = 1$ ms and a period of 10 ms. Express this signal as a cosine and sine function.

6. A voltage of $10 \text{ V} \cos(100t + 25°)$ is applied to a resistance of 25 Ω and an inductance of 0.50 H in parallel. What are the (a) phase angle and (b) rms value of the sum of the currents?

7. Draw the power triangle for the circuit in Prob. 6.

8. What is the capacitance, in units of farads, to completely correct the power factor of the circuit in Prob. 6, that is, to make $Q = 0$?

9. An 80 μF capacitor is in series with a 9 mH inductor. What is the total impedance (at 60 Hz), and is the circuit considered a leading or lagging circuit?

10. If the capacitor and inductor in Prob. 9 are connected in parallel, what is the total impedance, and is the circuit considered leading or lagging?

Circuit Theory

SOLUTIONS

1. As a function of time, t, with peak voltage V_p and angular frequency ω,

$$v(t) = V_p \sin(\omega t + \theta)$$

The peak voltage is proportional to the effective voltage, V_{eff},

$$V_p = V_{\text{eff}} \sqrt{2}$$

$$= 120 \text{ V}\sqrt{2}$$

$$= 169.71 \text{ V}$$

The angular frequency is

$$\omega = 2\pi f = 2\pi \left(60 \text{ s}^{-1}\right)$$

$$= 376.99 \text{ rad/s}$$

The given angle of 30° equals $\pi/6$ rad, thus

$$v(t) = \boxed{170 \sin\left(377t + \frac{\pi}{6}\right)}$$

2. (a) $15\cos(-180°) + j15\sin(-180°)$

$$= \boxed{-15 + j0}$$

(b) $10\cos 37° + j10\sin 37°$

$$= \boxed{7.99 + j6.02}$$

(c) $50\cos 120° + j50\sin 120°$

$$= \boxed{-25 + j43.3}$$

(d) $21\cos -90° + j21\sin -90°$

$$= \boxed{0 - j21}$$

3. (a) $\sqrt{(6)^2 + (7)^2}\angle \tan^{-1}\frac{7}{6}$

$$= \boxed{9.22\angle 49.4°}$$

(b) $\sqrt{(50)^2 + (-60)^2}\angle \tan^{-1}\frac{-60}{50}$

$$= \boxed{78.10\angle -50.19°}$$

(c) $\sqrt{(-75)^2 + (45)^2}\angle \tan^{-1}\frac{45}{-75}$

$$= \boxed{87.46\angle 149.04°}$$

(d) $\sqrt{(90)^2 + (-180)^2}\angle\tan^{-1}\left(\frac{-180}{90}\right)$

$$= \boxed{201.25\angle -63.43°}$$

4. (a) $\frac{1}{6}\sqrt{(7)^2 + (5)^2}\angle \tan^{-1}\frac{5}{7}$

$$= \boxed{1.4\angle 35.54°}$$

(b)

$$= \frac{21.4\angle 52.59°}{18.0\angle -33.69°}$$

$$= \boxed{1.19\angle 86.28°}$$

(c) $\boxed{0.588\angle 34°}$

5.

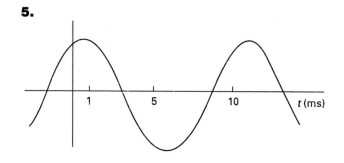

$$\omega = \frac{2\pi}{T} = \frac{2\pi}{0.01} = 200\pi \text{ rad/s}$$

$$10 \text{ ms} = 360°$$

$$1 \text{ ms} = 36°$$

For the sine, the positive zero crossing is at -1.5 ms, or $-54°$.

$$f = \boxed{F_p \sin(200\pi t + 54°) \text{ or } F_p \sin(200\pi t + 0.9)}$$

$$= \boxed{F_p \cos(200\pi t - 36°) \text{ or } F_p \cos(200\pi t - 0.6)}$$

6. Using a cosine reference,

(a)
$$I_R = \frac{10 \text{ V}\angle 25°}{25 \text{ } \Omega} = 0.4 \text{ A}\angle 25°$$

$$= 0.3625 + j0.1690 \text{ A}$$

$$I_L = \frac{10 \text{ V}\angle 25°}{(0.5)(100)\angle 90° \text{ } \Omega}$$

$$= 0.2 \text{ A}\angle -65°$$

$$= 0.0845 - j0.1813 \text{ A}$$

$$I_t = 0.4470 - j0.0123 \text{ A}$$

$$= 0.4472 \text{ A}\angle -1.6°$$

$$\text{phase angle} = \boxed{-1.6°}$$

(b) The calculations were made using peak values. Divide the total peak current by $\sqrt{2}$ to obtain the rms values.

$$I_{\text{rms}} = \frac{0.4472 \text{ A}}{\sqrt{2}} = \boxed{0.316 \text{ A}}$$

7. Use rms values. The magnitude, S, of the complex power vector is the product of the voltage, V, and the magnitude of the complex conjugate of the current, I^*.

$$S = VI^* = \frac{(10 \text{ V}\angle 25°)(0.4472\angle 1.6 \text{ A}°)}{\sqrt{2}\sqrt{2}} \text{ VA}$$

$$= 2.236 \text{ V}\angle 26.6°$$

$$= 2 + j1 \text{ VA}$$

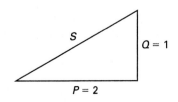

8. The capacitance current must cancel the inductance current, so its reactance must be $-j\omega L = -j50 \text{ } \Omega$.

$$X_C = \frac{1}{j\omega C} = \frac{-j}{100C} \rightarrow -j50 \text{ } \Omega$$

$$C = \frac{1}{(50 \text{ } \Omega)\left(100 \frac{\text{rad}}{\text{s}}\right)}$$

$$= \boxed{200 \times 10^{-6} \text{ F} \quad (200 \text{ } \mu\text{F})}$$

9. The impedence, Z_C, for the capacitor is

$$Z_C = \frac{1}{j\omega C} = \frac{1}{j\left(377 \frac{\text{rad}}{\text{s}}\right)(80 \times 10^{-6} \text{ F})}$$

$$= -j(33.16 \text{ } \Omega)$$

$$= 33.16 \text{ } \Omega\angle -90°$$

The impedence, Z_L, for the inductor is

$$Z_L = j\omega L = j\left(377 \frac{\text{rad}}{\text{s}}\right)(9 \times 10^{-3} \text{ H})$$

$$= j3.39 \text{ } \Omega$$

$$= 3.39 \text{ } \Omega\angle 90°$$

For a series circuit, the impedance is

$$Z_{\text{total}} = Z_C + Z_L$$

$$= -j33.16 \text{ } \Omega + j3.39 \text{ } \Omega$$

$$= -j29.77 \text{ } \Omega$$

$$= \boxed{29.77 \text{ } \Omega\angle -90°}$$

The impedance angle is negative. The circuit is capacitative and therefore leading.

10. For a parallel circuit, the impedance is

$$\frac{1}{Z_{\text{total}}} = \frac{1}{Z_C} + \frac{1}{Z_L}$$

$$= \frac{1}{33.16 \text{ } \Omega\angle -90°} + \frac{1}{3.39 \text{ } \Omega\angle 90°}$$

$$= 0.0302 \text{ } \Omega\angle 90° + 0.2950 \text{ } \Omega\angle -90°$$

$$= j0.0302 \text{ } \Omega - j0.2950 \text{ } \Omega$$

$$= -j0.2648 \text{ } \Omega$$

Therefore,

$$Z_{\text{total}} = \frac{1}{-j0.2648 \ \Omega} = \frac{j}{0.2648 \ \Omega}$$

$$= \boxed{3.78 \ \Omega \angle 90°}$$

The impedance angle is positive. The circuit is inductive and therefore lagging.

28 Transformers

PRACTICE PROBLEMS

1. Which of the following describes the leakage flux?
- (A) flux outside the core
- (B) flux linking the primary
- (C) flux linking the secondary
- (D) flux other than the mutual flux

2. For the transformer shown, which of the following statements is true?

- (A) $M > 0$
- (B) $M < 0$
- (C) $X_m > 0$ and $X_p > 0$
- (D) Both (A) and (C) are true.

3. An ideal transformer has 90 turns on the primary and 2200 turns on the secondary. The load draws 8 A at 0.8 pf. What is the current in the primary?
- (A) 90 A $\angle 37°$
- (B) 196 A $\angle 37°$
- (C) 320 A $\angle 1°$
- (D) 333 A $\angle 0.2\pi$

4. An ideal transformer with a 120 V, 60 Hz source has a turns ratio of 10. What is the peak flux in the secondary?

5. A step-down transformer consists of 200 primary turns and 40 secondary turns. The primary voltage is 550 V. If an impedance of 4.2 Ω is the secondary load, what are the (a) secondary voltage, (b) primary current, and (c) secondary current?

6. For the circuit shown, if the amplifier and load impedances are matched, what is the turns ratio?

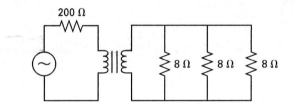

7. A 100 V (rms) source with an internal resistance of 8 Ω is connected to a 200/100-turn step-down transformer. An impedance of 6 $\Omega + j8$ is connected to the secondary. Find the (a) primary current and (b) secondary power.

SOLUTIONS

1. Leakage flux is defined as flux that goes beyond its intended path and does not serve its intended purpose—that is, the flux that links only one coil or the other, but not both, in a two-coil transformer. The flux linking both coils is the mutual flux.

The answer is (D).

2. Both assumed currents enter at the dotted terminals. Thus, the mutual inductor is positive. Further, the mutual reactance, X_m, and the self-inductance of the coils, X_p (or X_s), are the same sign.

The answer is (D).

3. The turns ratio a for a transformer with N windings and currents I (p, primary; s, secondary) is

$$a = \frac{N_p}{N_s} = \frac{I_s}{I_p}$$

Therefore,

$$a = \frac{N_p}{N_s} = \frac{90}{2200} = 0.04091$$

$$= \frac{I_s}{I_p}$$

The magnitude of the current is

$$I_p = \frac{I_s}{a} = \frac{8 \text{ A}}{0.04091} = 195.6 \text{ A}$$

In an ideal transformer, no phase shift occurs, thus the power factor is

$$\text{pf} = \cos\psi$$

$$0.8 = \cos\psi$$

$$\psi = \cos^{-1} 0.8 = 36.87°$$

Thus,

$$\boxed{\mathbf{I}_p = 196 \text{ A}\angle 37°}$$

The answer is (B).

4. The effective voltage, $V_s(t)$, induced in the secondary within a transformer with mutual flux Φ_m is

$$v_s(t) = -4.44\, N_s f \Phi_m \cos\omega t$$

Because only the peak flux was requested,

$$V_s = 4.44\, N_s f \Phi_m$$

$$\Phi_m = \frac{V_s}{4.44\, N_s f}$$

Now, from the turns ratio,

$$a = \frac{N_p}{N_s} = \frac{V_p}{V_s}$$

$$V_s = \left(\frac{N_s}{N_p}\right) V_p = \tfrac{1}{10} (120 \text{ V}) = 12 \text{ V}$$

Substituting gives

$$\Phi_m = \frac{V_s}{4.44\, N_s f} = \frac{12 \text{ V}}{(4.44)(1)(60 \text{ s}^{-1})}$$

$$= \boxed{4.54 \times 10^{-2} \text{ Wb } (45 \text{ mWb})}$$

Note that this is the same flux present in the primary.

5. (a) $\quad V_s = \left(\dfrac{N_s}{N_p}\right) V_p = \left(\dfrac{40}{200}\right)(550 \text{ V})$

$$= \boxed{110 \text{ V}}$$

(b) $\quad I_p = \dfrac{V_s^2}{V_p Z_s} = \dfrac{(110 \text{ V})^2}{(550 \text{ V})(4.2 \text{ }\Omega)}$

$$= \boxed{5.24 \text{ A}}$$

(c) $\quad I_s = \dfrac{V_s}{Z_s} = \dfrac{110 \text{ V}}{4.2 \text{ }\Omega}$

$$= \boxed{26.2 \text{ A}}$$

6. $\quad \dfrac{N_p}{N_s} = \sqrt{\dfrac{Z_p}{Z_s}} = \sqrt{\dfrac{200 \text{ }\Omega}{\dfrac{8 \text{ }\Omega}{3}}}$

$$= \boxed{8.66}$$

7. Consider the following circuit.

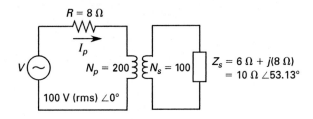

The circuit can be simplified to

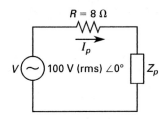

$$\mathbf{Z}_p = \left(\frac{N_p}{N_s}\right)^2 \mathbf{Z}_s = \left(\frac{200}{100}\right)^2 (10 \ \Omega\angle 53.13°)$$

$$= 40 \ \Omega\angle 53.13°$$
$$= (24 + j32) \ \Omega$$

The circuit can be further simplified to

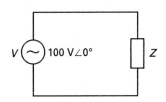

$$Z = \sqrt{(\Sigma R)^2 + (\Sigma X_L - \Sigma X_C)^2}$$

$$= \sqrt{(R + R_{Z_m})^2 + (X_{Z_m})^2}$$

$$= \sqrt{(8 \ \Omega + 24 \ \Omega)^2 + (32 \ \Omega)^2}$$

$$= 45.25 \ \Omega$$

$$\phi\mathbf{z} = \tan^{-1} \frac{\Sigma X_L - \Sigma X_C}{\Sigma R}$$

$$= \tan^{-1} \frac{X_{Z_m}}{R + R_{Z_m}}$$

$$= \tan^{-1} \frac{32 \ \Omega}{8 \ \Omega + 24 \ \Omega}$$

$$= 45°$$

$$\mathbf{Z} = 45.25 \ \Omega\angle 45°$$

(a) $\quad \mathbf{I}_p = \dfrac{\mathbf{V}}{\mathbf{Z}} = \dfrac{100 \ \text{V}\angle 0°}{45.25 \ \Omega\angle 45°}$

$$= \boxed{2.21 \ \text{A}_{\text{rms}}\angle -45°}$$

(b) $\quad P_s = I_s^2 R_s = I_s^2 Z_s \cos \phi$

$$= \left(\frac{N_p I_p}{N_s}\right)^2 Z_s \cos \phi$$

$$= \left(\frac{(200)(2.21 \ \text{A})}{100}\right)^2 (10 \ \Omega) \cos 53.13°$$

$$= \boxed{117.2 \ \text{W}}$$

Circuit Theory

29 Linear Circuit Analysis

PRACTICE PROBLEMS

1. For which of the following circuit elements is super-position valid?
- (A) independent source
- (B) dependent source
- (C) charged capacitor
- (D) inductor with $\mathbf{I}_{initial} \neq 0$

2. A 5 Ω length of copper wire is used in a circuit designed for a maximum temperature of 50°C. What is the resistance of the wire at maximum design temperature specification?

3. Two square parallel plates (0.04 m × 0.04 m) are separated by a 0.1 cm thick insulator with a dielectric constant of 3.4. The plates are connected across 200 V. What is the capacitance?

4. For the parallel plates described in Prob. 3, what charge exists on the plates?

5. An 8 mH inductor carries 15 A of current. What is the average magnetic flux present?

6. For the circuit shown, which resistor(s) is (are) redundant?

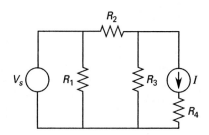

- (A) R_1
- (B) R_1 and R_3
- (C) R_2 and R_4
- (D) R_1 and R_4

7. For the circuit shown, what current is flowing in the 8 Ω resistor?

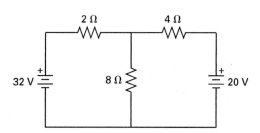

8. For the circuit shown, what current is flowing in the 6 Ω resistor?

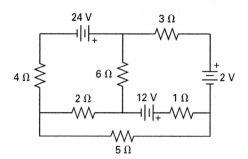

9. At $t = 0$, the switch connects to position A. At $t = 5 \times 10^{-4}$ s, the switch is moved to position B. Assume there is no current anywhere in the circuit prior to $t = 0$. (a) What is the current flowing in the circuit just prior to the switch moving to position B? (b) What is the equation of the current after the switch is moved to position B?

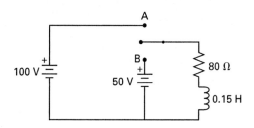

10. For the circuit shown, find the impedance, z, parameters.

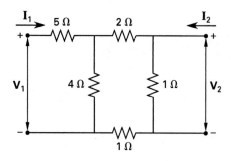

SOLUTIONS

1. Capacitors and inductors are linear elements for which superposition would be valid only if they were initially uncharged or without an initial magnetic field, respectively. Therefore, neither (C) nor (D) is the answer. The output of dependent sources depends on electrical parameters elsewhere in the circuit and is therefore not linear. Therefore, (B) is not the answer.

Independent sources are linear, and superposition can be applied.

The answer is (A).

2. Resistance is a function of the resistivity, ρ, the length l, and the area A.

$$R = \frac{\rho l}{A}$$

The length and cross-sectional area are unknown. Therefore, use the following ratio.

$$\frac{R_{50}}{R_{20}} = \frac{\rho_{50}\dfrac{l}{A}}{\rho_{20}\dfrac{l}{A}} = \frac{\rho_{50}}{\rho_{20}}$$

Thus,

$$R_{50} = R_{20}\frac{\rho_{50}}{\rho_{20}}$$

Solving for ρ_{50} gives

$$\rho_{50} = \rho_{20}\left(1 + \alpha_{20}\left(T - 20°\mathrm{C}\right)\right)$$

$$= \left(1.8 \times 10^{-8}\ \Omega\cdot\mathrm{m}\right)$$

$$\times \left(1 + (3.9 \times 10^{-3°}\mathrm{C}^{-1})(50°\mathrm{C} - 20°\mathrm{C})\right)$$

$$= 20.11 \times 10^{-9}\ \Omega\cdot\mathrm{m}$$

Substituting the given and calculated information gives

$$R_{50} = R_{20}\frac{\rho_{50}}{\rho_{20}} = (5\ \Omega)\left(\frac{20.11 \times 10^{-9}\ \Omega\cdot\mathrm{m}}{1.8 \times 10^{-8}\ \Omega\cdot\mathrm{m}}\right)$$

$$= \boxed{5.6\ \Omega}$$

3. The capacitance is a function of the permittivity $\varepsilon_r\varepsilon_0$, the area of the plates, A, and the distance, r, between them.

$$C = \frac{\varepsilon A}{r} = \frac{\varepsilon_r\varepsilon_0 A}{r}$$

$$= \frac{(3.4)\left(8.854 \times 10^{-12}\dfrac{\mathrm{C}^2}{\mathrm{N}\cdot\mathrm{m}^2}\right)(0.04\ \mathrm{m})^2}{0.001\ \mathrm{m}}$$

$$= \boxed{4.817 \times 10^{-11}\ \mathrm{F}}$$

4. The charge on the plates is

$$Q = CV_{\text{plates}} = (4.817 \times 10^{-11} \text{ F})(200 \text{ V})$$

$$\boxed{= 9.63 \times 10^{-9} \text{ C}}$$

5. The average energy in the inductor's magnetic field depends on the self-inductance, L, and the effective current, I.

$$U = \tfrac{1}{2}LI^2 = \left(\tfrac{1}{2}\right)\left(8 \times 10^{-3} \text{ H}\right)(15 \text{ A})^2$$
$$= 0.90 \text{ J}$$

The energy in terms of the magnetic flux, Ψ, is

$$U = \tfrac{1}{2}\left(\frac{\Psi^2}{L}\right)$$

$$\Psi = \sqrt{2UL} = \sqrt{(2)(0.90 \text{ J})(8 \times 10^{-3} \text{ H})}$$

$$\boxed{= 0.120 \text{ Wb}}$$

6. Resistors in parallel with voltage sources, or in series with current sources, do not affect the remainder of the circuit. Thus, R_1 and R_4 are redundant.

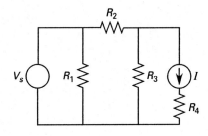

The answer is (D).

7. The circuit can be represented as

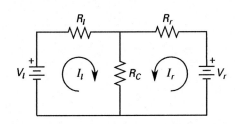

From the loop-current method,

$$V_l = R_l I_l + R_c(I_l + I_r)$$

$$V_r = R_r I_r + R_c(I_l + I_r)$$

$$I_l + I_r = \frac{R_r V_l + R_l V_r}{R_r R_l + R_r R_c + R_l R_c}$$

$$= \frac{(4 \text{ }\Omega)(32 \text{ V}) + (2 \text{ }\Omega)(20 \text{ V})}{(4 \text{ }\Omega)(2 \text{ }\Omega) + (4 \text{ }\Omega)(8 \text{ }\Omega) + (2 \text{ }\Omega)(8 \text{ }\Omega)}$$

$$\boxed{= 3 \text{ A} \quad [\text{down}]}$$

If the first two equations are solved simultaneously,

$$I_l = 4 \text{ A}$$

$$I_r = -1 \text{ A}$$

8. The circuit can be represented as

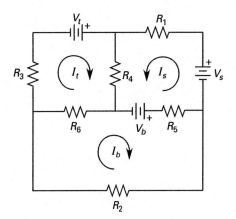

Using the loop-current method,

$$V_t = R_4(I_t + I_s) + R_6(I_t - I_b) + R_3 I_t$$

$$V_s + V_b = R_1 I_s + R_4(I_t + I_s) + R_5(I_s + I_b)$$

$$V_b + R_6(I_t - I_b) = R_5(I_b + I_s) + R_2 I_b$$

After rearranging, substituting in the constants, and evaluating the coefficients, the previous three equations are

$$24 \text{ V} = (12 \text{ }\Omega)I_t + (6 \text{ }\Omega)I_s - (2 \text{ }\Omega)I_b$$

$$14 \text{ V} = (6 \text{ }\Omega)I_t + (10 \text{ }\Omega)I_s + (1 \text{ }\Omega)I_b$$

$$12 \text{ V} = -(2 \text{ }\Omega)I_t + (1 \text{ }\Omega)I_s + (8 \text{ }\Omega)I_b$$

Solving simultaneously,

$$I_t = 2.53 \text{ A}$$

$$I_s = -0.336 \text{ A}$$

$$I_t + I_s = \boxed{2.194 \text{ A} \quad [\text{down}]}$$

9. (a) Before the switch is moved to position B,

$$I(t) = \left(\frac{V_{100}}{R}\right)\left(1 - e^{\frac{-Rt}{L}}\right)$$

$$I\left(5 \times 10^{-4}\right) = \left(\frac{100 \text{ V}}{80 \ \Omega}\right)\left(1 - e^{\frac{-(80 \ \Omega)(5 \times 10^{-4} \text{ s})}{0.15 \text{ H}}}\right)$$

$$= \boxed{0.2926 \text{ A} \quad [\text{clockwise}]}$$

(b) Move the switch to position B. Solve for currents I_1 and I_2, whose sum is I_N, the current flowing in the circuit.

$$I_1(t) = I_0 e^{\frac{-Rt}{L}} \quad [\text{clockwise}]$$

$$I_2(t) = \left(\frac{V_{50}}{R}\right)\left(1 - e^{\frac{-Rt}{L}}\right) \quad [\text{clockwise}]$$

$$I_N(t) = I_1(t) + I_2(t)$$

$$= I_0 e^{\frac{-Rt}{L}} + \left(\frac{V_{50}}{R}\right)\left(1 - e^{\frac{-Rt}{L}}\right)$$

$$= \frac{50 \text{ V}}{80 \ \Omega} + \left(0.2926 \text{ A} - \frac{50 \text{ V}}{80 \ \Omega}\right)e^{\frac{-(80 \ \Omega)t}{0.15 \text{ H}}}$$

$$= \boxed{0.625 - 0.332 e^{-(533s^{-1})t} \text{ A} \quad [\text{clockwise}]}$$

10. Consider the circuit shown.

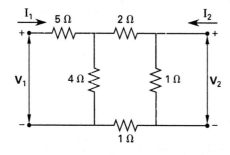

Because all the impedances are resistors, the phasor notation is dropped for convenience. From the deriving equations, the impedance is

$$z_{11} = \left.\frac{V_1}{I_1}\right|_{I_2=0 \text{ A}}$$

The circuit, redrawn with $I_2 = 0$ A, is

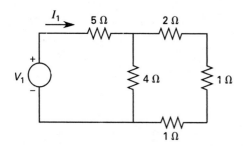

The equivalent impedance looking to the right, into the input terminals, is

$$z_e = R_e = 5 \ \Omega + \frac{(4 \ \Omega)(2 \ \Omega + 1 \ \Omega + 1 \ \Omega)}{4 \ \Omega + 2 \ \Omega + 1 \ \Omega + 1 \ \Omega}$$

$$= 7 \ \Omega$$

Redrawing the circuit again gives

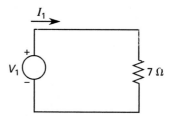

Using Ohm's law gives

$$V_1 = I_1(7 \ \Omega)$$

$$\frac{V_1}{I_1} = 7 \ \Omega$$

Thus,

$$z_{11} = \boxed{7 \ \Omega}$$

From the deriving equations, the impedance is

$$z_{21} = \left.\frac{V_2}{I_1}\right|_{I_2=0 \text{ A}}$$

The redrawn circuit is identical to that for z_{11}, since $I_2 = 0$ A.

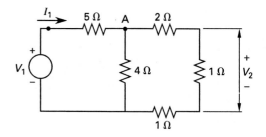

Using the concept of a current divider at node A gives the following for the current through the branch containing V_2.

$$I_{V_2} = I_1\left(\frac{4 \ \Omega}{4 \ \Omega + (2 \ \Omega + 1 \ \Omega + 1 \ \Omega)}\right) = \tfrac{1}{2}I_1$$

Apply Ohm's law across the 1 Ω resistor associated with V_2, using the calculated current.

$$V_2 = \tfrac{1}{2}I_1(1 \ \Omega) = \tfrac{1}{2}I_1$$

$$\frac{V_2}{I_1} = \tfrac{1}{2} \ \Omega$$

Thus,

$$z_{21} = \left.\frac{V_2}{I_1}\right|_{I_2 = 0 \text{ A}} = \tfrac{1}{2}\ \Omega$$

From the deriving equations,

$$z_{12} = \left.\frac{V_1}{I_2}\right|_{I_1 = 0 \text{ A}}$$

The circuit, redrawn with $I_1 = 0$ A, is

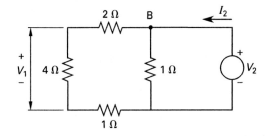

Using a current divider at node B gives the following for the branch current flowing through the 4 Ω resistor associated with V_1.

$$I_{V_1} = I_2 \left(\frac{1\ \Omega}{1\ \Omega + 2\ \Omega + 4\ \Omega + 1\ \Omega}\right) = \tfrac{1}{8}I_2$$

Applying Ohm's law for the 4 Ω resistor gives

$$V_1 = \tfrac{1}{8}I_2\,(4\ \Omega) = \tfrac{1}{2}I_2$$

$$\frac{V_1}{I_2} = \tfrac{1}{2}\ \Omega$$

Thus,

$$z_{12} = \left.\frac{V_1}{I_2}\right|_{I_1 = 0 \text{ A}} = \tfrac{1}{2}\ \Omega$$

From the deriving equations,

$$z_{22} = \left.\frac{V_2}{I_2}\right|_{I_1 = 0 \text{ A}}$$

The redrawn circuit is identical to that for z_{12}, since $I_1 = 0$ A.

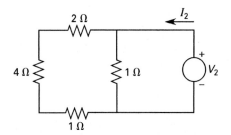

The equivalent impedance looking to the left, into the output terminals, is

$$z_e = R_e = \frac{(1\ \Omega)\,(2\ \Omega + 4\ \Omega + 1\ \Omega)}{1\ \Omega + (2\ \Omega + 4\ \Omega + 1\ \Omega)} = \tfrac{7}{8}\ \Omega$$

Redrawing the circuit again gives

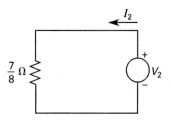

Using Ohm's law gives

$$V_2 = I_2\left(\tfrac{7}{8}\ \Omega\right)$$

$$\frac{V_2}{I_2} = \tfrac{7}{8}\ \Omega$$

Thus,

$$z_{22} = \tfrac{7}{8}\ \Omega$$

30 Transient Analysis

PRACTICE PROBLEMS

1. Consider the circuit shown. What is the time constant?

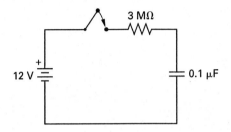

(A) 1.0 μs
(B) 36 μs
(C) 0.30 s
(D) 3.0 s

2. With an initially uncharged capacitor in the circuit in Prob. 1, what is the initial current?

(A) 0.0 A
(B) 4.0 μA
(C) 4.0 A
(D) 120 A

3. How long does it take the circuit shown in Prob. 1 to reach steady-state?

4. What is the rise time of the circuit in Prob. 1?

5. Consider the circuit shown. Write the equations for loops 1 and 2.

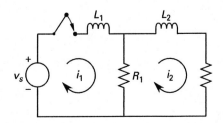

6. A series RC circuit contains a 4 Ω resistor and an uncharged 300 μF capacitor. The switch is closed at $t = 0$ s, connecting the circuit across a 500 V source. For the circuit shown, determine the current as a function of time.

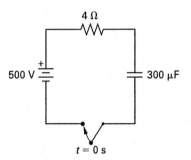

7. The switch in the circuit shown is closed at $t = 0$ s and opened again at $t = 0.05$ s. How much energy is stored in the initially uncharged capacitor?

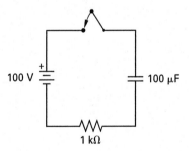

8. Find the current through the 80 Ω resistor (a) immediately after the switch is closed and (b) 2 s after the switch is closed. The switch has been open for a long time.

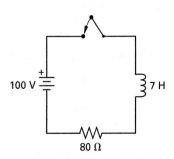

9. For the circuit shown, determine the steady-state inductance current, i_L, (a) with the switch closed and (b) with the switch open.

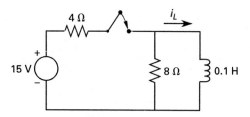

10. For the circuit shown, determine the steady-state capacitance voltage, V_C, (a) with the switch closed, and (b) with the switch open.

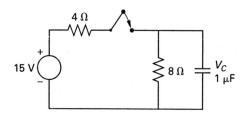

11. For the circuit shown, determine the steady-state inductance current, i_L, (a) with the switch closed and (b) with the switch open.

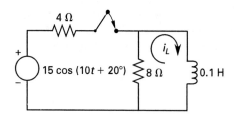

12. The circuit in Prob. 10 has had the switch closed for a long time. At $t = 0$ s, the switch opens. Determine the capacitance voltage, $v_C(t)$, for $t > 0$ s, for a capacitance of 1×10^{-6} F.

13. The circuit in Prob. 10 has had the switch open for a long time. At $t = 0$ s, the switch closes. Determine the capacitance voltage, $v_C(t)$, for $t > 0$ s.

14. A system is described by the equation

$$12 \sin 2t = \frac{d^2v}{dt^2} + 5\frac{dv}{dt} + 4v$$

The initial conditions on the voltage are

$$v(0) = -5 \text{ V}$$

$$\frac{dv(0)}{dt} = 2 \text{ V/s}$$

Determine $v(t)$ for $t > 0$ s.

15. For the circuit shown, the current source is

$$i_s(t) = 0.01 \cos 500t$$

Find the transfer function.

$$G(s) = \frac{V_C(s)}{I_s(s)}$$

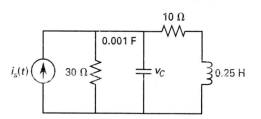

16. For a series RLC circuit of 10 Ω, 0.5 H, and 100 μF, determine the (a) resonant frequency, (b) quality factor, and (c) bandwidth.

17. For a parallel RLC circuit of 10 Ω, 0.5 H, and 100 μF, determine the (a) resonant frequency, (b) quality factor, and (c) bandwidth.

18. The switch of the circuit shown has been open for a long time and is closed at $t = 0$ s. Determine the currents flowing in the two capacitances at the instant the switch closes. Give the magnitudes and directions.

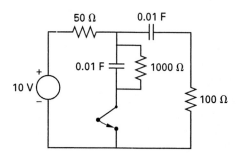

19. A system is described by the following differential equation.

$$20 \sin 4t = \frac{d^2i}{dt^2} + 4\frac{di}{dt} + 4i$$

The initial conditions are

$$i(0) = 0 \text{ A}$$

$$\frac{di(0)}{dt} = 4$$

Determine $i(t)$ for $t > 0$ s.

20. The circuit shown has the 1 μF capacitance charged to 100 V and the 2 μF capacitance uncharged before the switch is closed. Determine the (a) energy stored in each capacitance at the instant the switch closes, (b) energy stored in each capacitance a long time after the switch closes, and (c) energy dissipated in the resistance.

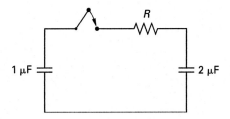

SOLUTIONS

1.

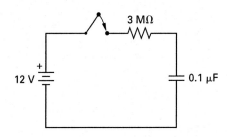

The time constant is

$$\tau = RC = \left(3 \times 10^6 \ \Omega\right)\left(0.1 \times 10^{-6} \ \text{F}\right)$$

$$= \boxed{0.3 \ \text{s}}$$

The answer is (C).

2. The capacitor initially acts as a short. Thus, from Ohm's law,

$$V = IR$$

$$I = \frac{V}{R} = \frac{12 \ \text{V}}{3 \times 10^6 \ \Omega}$$

$$= \boxed{4.0 \times 10^{-6} \ \text{A} \quad (4.0 \ \mu\text{A})}$$

The answer is (B).

3. Steady-state is reached in approximately 5 time constants. Thus,

$$t = 5\tau$$

The time constant is $RC = 0.3$ s. Substituting,

$$t = (5)(0.3 \ \text{s}) = \boxed{1.5 \ \text{s}}$$

4. The rise time is most often determined by measurements in the lab. Nevertheless, the rise time is

$$t_r = 2.2\tau$$

The time constant was found in Prob. 1. Substituting gives

$$t_r = (2.2)(0.3 \ \text{s}) = \boxed{0.660 \ \text{s}}$$

Circuit Theory

5.

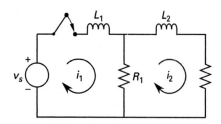

Kirchoff's voltage law (KVL) for loop 1 gives

$$v_s - L_1 \left(\frac{di_1}{dt} \right)$$

$$- (i_1 - i_2)R_1 = 0$$

$$\boxed{ v_s = L_1 \left(\frac{di_1}{dt} \right) + (i_1 - i_2)R_1 }$$

KVL for loop 2 gives

$$-(i_2 - i_1)R_1 - L_2 \left(\frac{di_2}{dt} \right) - i_2 R_2 = 0$$

$$-i_2 R_1 + i_1 R_1 - L_2 \left(\frac{di_2}{dt} \right) - i_2 R_2 = 0$$

$$\boxed{ L_2 \left(\frac{di_2}{d_t} \right) + R_1(i_2 - i_1) + i_2 R_2 = 0 }$$

6. From Table 30.1, the current response for a series RC circuit is

$$i(t) = \left(\frac{V_{\text{bat}} - V_0}{R} \right) e^{\frac{-t}{\tau}} = \left(\frac{500 \text{ V} - 0 \text{ V}}{4 \ \Omega} \right) e^{\frac{-t}{1.2 \times 10^{-3}}}$$

$$= \boxed{ 125 e^{-833t} \text{ A} }$$

7.

$$Q(t) = CV_C(t) = CV \left(1 - e^{\frac{-t}{RC}} \right)$$

$$E_C(t) = \tfrac{1}{2} CV_C^2(t) = \tfrac{1}{2} CV^2 \left(1 - e^{\frac{-t}{RC}} \right)^2$$

$$E_C(0.05) = \left(\tfrac{1}{2} \right) (100 \times 10^{-6} \text{ F})(100 \text{ V})^2$$

$$\times \left(1 - e^{\frac{-0.05 \text{ s}}{(10^3 \ \Omega)(100 \times 10^{-6} \text{ F})}} \right)^2$$

$$= \boxed{ 0.0774 \text{ J} }$$

8. (a) Since the current in the inductor cannot change instantaneously,

$$I(0) = \boxed{ 0 \text{ A} }$$

(b) $$I(t) = \left(\frac{V}{R} \right) \left(1 - e^{\frac{-Rt}{L}} \right)$$

$$I(2) = \left(\frac{100 \text{ V}}{80 \ \Omega} \right) \left(1 - e^{\frac{-(80 \ \Omega)(2 \text{ s})}{7 \text{ H}}} \right)$$

$$= \boxed{ 1.25 \text{ A} \qquad \text{[down]} }$$

9. (a) Use KVL around the outer loop. With the switch closed, KVL gives

$$15 = 4i_L + v_L$$

$$v_L = L \left(\frac{di_L}{dt} \right) = 0 \text{ V} \quad \text{[in steady-state, } ss\text{]}$$

$$i_{L_{ss}} = i_{ss} = \frac{15 \text{ A}}{4 \text{ A}}$$

The current through the 8 Ω resistor is $v_L/8 = 0$ V, therefore all of i_{ss} flows through the inductance in steady-state.

(b) With the switch open, the inductor dissipates its energy in the 8 Ω resistor.

$$i_{L_{ss}} = \boxed{ i_{ss} = 0 \text{ A} }$$

10. (a) With the switch closed,

$$i_C = C \left(\frac{dv_C}{dt} \right)$$

In steady-state,

$$\frac{dv_C}{dt} = 0$$

Therefore, $i_C \to 0$. By voltage division,

$$v_C = \left(\frac{8 \ \Omega}{12 \ \Omega} \right) (15 \text{ V}) = \boxed{ 10 \text{ V} }$$

With the switch open, the capacitor dissipates its energy in the 8 Ω resistor.

(b)

$$\boxed{ v_C = 0 \text{ V} }$$

11. (a) A Thevenin equivalent circuit, seen from the inductance, is obtained.

$$v_{oc} = v_{Th} = \left(\frac{8\ \Omega}{12\ \Omega}\right) v_s = 10\ \text{V} \cos(10t + 20°)$$

$$i_{sc} = \frac{15 \cos(10t + 20°)\ \text{V}}{4\ \Omega} = 3.75 \cos(10t + 20°)\ \text{A}$$

$$R_{Th} = \frac{v_{oc}}{i_{sc}} = \frac{10\ \text{V}}{3.75\ \text{A}} = 8/3\ \Omega$$

Taking $\cos(10t + 20°)$ as the phasor reference and using the maximum value, for convenience, instead of the rms value, gives

$$V_{Th} = 10\ \text{V}\angle 0°$$

$$Z_L = j\omega L = \left(j10\ \frac{\text{rad}}{\text{s}}\right)(0.1\ \text{H}) = j1\ \Omega$$

$$i_L = \frac{V_{Th}}{R_{Th} + j1\ \Omega}$$

$$= \left(\frac{10\ \text{V}}{\frac{8}{3}\ \Omega + j1\ \Omega}\right)\left(\frac{\frac{8}{3}\ \Omega - j1\ \Omega}{\frac{8}{3}\ \Omega - j1\ \Omega}\right)$$

$$= \left(\tfrac{90}{73}\ \text{A}\right)\left(\tfrac{8}{3} - j1\right) = \boxed{3.51\ \text{A}\angle -20.6°}$$

As a time function,

$$i_L(\infty) = 3.51 \cos(10t + 20° - 20.6°)$$

$$= 3.51 \cos(10t - 0.6°)\ \text{A}$$

(b) With the switch open,

$$i_L(\infty) = \boxed{0\ \text{A}}$$

12. With the switch closed for a long time, the capacitance voltage is 10 V (see Prob. 10). The initial voltage when the switch opens is therefore 10 V.

With the switch open, KVL is

$$v_C = 8i$$

$$i_C = 10^{-6}\left(\frac{dv_C}{dt}\right)$$

$$i_C = -i$$

$$v_C + (8 \times 10^{-6})\left(\frac{dv_C}{dt}\right) = 0\ \text{V}$$

$$\frac{dv_C}{dt} + (1.25 \times 10^5)v_C = 0\ \text{V}$$

Any appropriate mathematical method may now be used to solve the differential equation.

Let $\alpha = 1.25 \times 10^5$. The solution is of the form

$$v_C = V(t_1)e^{-\alpha(t-t_1)}$$

Taking $t_1 = 0$ s, $v_C(0) = 10$ V.

$$\frac{dv_C}{dt} = -\alpha v_C$$

$$\alpha = 1.25 \times 10^5$$

$$v_C(t) = \boxed{10e^{-(1.25\times 10^5)t}\ \text{V}}$$

13. The initial capacitance voltage is zero. Kirchhoff's current law (KCL) at the upper node, with the lower node as a reference, is

$$\frac{15\ \text{V} - v_C}{4\ \Omega} = \frac{v_C}{8\ \Omega} + (10^{-6}\ \text{F})\left(\frac{dv_C}{dt}\right)$$

This is manipulated to

$$\left(\tfrac{8}{3} \times 10^{-6}\right)\left(\frac{dv_C}{dt}\right) + v_C = 10\ \text{V}$$

$$V_{ss} = 10\ \text{V}$$

$$\tau = \tfrac{8}{3} \times 10^{-6}\ \text{s}\quad (2.67 \times 10^{-6}\ \text{s})$$

$$V_0 = 0\ \text{V}$$

$$v_C(t) = V_{ss} + (V_0 - V_{ss})e^{\frac{-t}{\tau}}$$

$$= \boxed{(10)\left(1 - e^{-(3.75\times 10^5)t}\right)u(t)\ \text{V}}$$

14. The homogeneous differential equation is

$$\frac{d^2 v}{dt^2} + 5\frac{dv}{dt} + 4v = 0\ \text{V}$$

This has solutions e^{st},

$$\frac{d^2 v}{dt^2} \rightarrow s^2 V$$

$$\frac{dv}{dt} \rightarrow sV$$

$$s^2 + 5s + 4 = 0$$

$$(s+4)(s+1) = 0$$

Transient solutions are of the form

$$Ae^{-4t} + Be^{-t}$$

The steady-state sinusoidal solution must be found before A and B can be evaluated.

The phasor solution is

$$12\sin 2t \Leftrightarrow 12 \text{ V}\angle 0°$$

$$\omega = 2$$

$$\frac{d}{dt} \Rightarrow j\omega = j2$$

$$\frac{d^2}{dt^2} \Rightarrow -\omega^2 = -4$$

The phasor equation is

$$12 \text{ V} = -4V_{ss} + (5)(j2)V_{ss} + 4V_{ss}$$

$$= j10V_{ss}$$

$$V_{ss} = \frac{1.2}{j} = 1.2 \text{ V}\angle -90°$$

$$v_{ss} = 1.2 \text{ V}\sin(2t - 90°)$$

$$= -1.2 \text{ V}\cos 2t$$

$$v(t) = Ae^{-4t} + Be^{-t} - 1.2\cos 2t$$

$$v(0) = A + B - 1.2 \text{ V} = -5 \text{ V}$$

$$A + B = -3.8 \text{ V}$$

$$\frac{dv}{dt} = -4Ae^{-4t} - Be^{-t} + 2.4\sin 2t$$

$$\frac{dv}{dt}(0) = -4A - B = 2$$

$$A = 0.6 \text{ V}$$

$$B = -4.4 \text{ V}$$

$$v(t) = \boxed{(0.6e^{-4t} - 4.4e^{-t} - 1.2\cos 2t)u(t) \text{ V}}$$

15. The impedance of the inductance branch is

$$Z_{LR} = 10 + 0.25s \ \Omega$$

KCL applied to the upper node is

$$I_s = \frac{V_C}{30 \ \Omega} + (0.001 \text{ F})sV_C + \frac{V_C}{10 \ \Omega + (0.25 \text{ H})s}$$

Solving,

$$G = \frac{V_C}{I_s} = \left(\frac{3 \times 10^4 + 750s}{0.75s^2 + 55s + 4000}\right) \ \Omega$$

$$= \boxed{(1000)\left(\frac{s + 40}{s^2 + 73.33s + 5333}\right) \ \Omega}$$

16. (a) $\quad Z = 10 \ \Omega + 0.5s \ \Omega + \dfrac{10^4}{s} \ \Omega$

$$= \left(\frac{1}{2}\right)\left(\frac{s^2 + 20s + 2 \times 10^4}{s}\right) \ \Omega$$

$$= \left(\frac{1}{2s}\right)\left(s^2 + \frac{\omega_o}{Q}s + \omega_o^2\right) \ \Omega$$

The resonant frequency is

$$\omega_o = \boxed{\sqrt{2} \times 10^2 \text{ rad/s}}$$

(b) The quality factor is

$$Q = \frac{\omega_o}{20} = \boxed{5\sqrt{2}}$$

(c) The bandwidth is

$$\text{BW} = \frac{\omega_o}{Q} = \boxed{20 \text{ rad/s}}$$

17. (a) $\quad Y = 0.1 + \dfrac{1}{0.5s} + 10^{-4}s$

$$= \left(\frac{10^{-4}}{s}\right)(s^2 + 10^3 s + 2 \times 10^4)$$

$$= \left(\frac{10^{-4}}{s}\right)\left(s^2 + \frac{\omega_0}{Q}s + \omega_o^2\right)$$

The resonant frequency is

$$\omega_0 = \sqrt{2 \times 10^4} = \boxed{100\sqrt{2} \text{ rad/s}}$$

(b) The quality factor is

$$Q = \frac{\omega_0}{\text{BW}} = \frac{\sqrt{2}}{10} = \boxed{0.1\sqrt{2}}$$

(c) The bandwidth is

$$\text{BW} = \frac{\omega_0}{Q} = \boxed{10^3 \text{ rad/s}}$$

18. With the switch open for a long time, the upper-right capacitor is charged to 10 V and the lower (center) capacitor is discharged to 0 V. At the instant of switching, the capacitors can be modeled as ideal voltage sources with their initial voltages. The resulting equivalent circuit is shown.

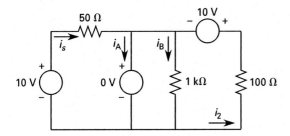

KVL on the left loop is

$$10 \text{ V} - (50 \ \Omega)i_s + 0.0 = 0$$

$$i_s = 1/5 \text{ A}$$

KVL on the right loop is

$$0.0 + 10 \text{ V} - (100 \ \Omega)i_2 = 0$$

$$i_2 = 1/10 \text{ A}$$

The current in the right (upper) capacitor is i_2, with 0.1 A flowing left.

As the voltage across the 1 kΩ resistor is zero, $i_B = 0$ A. By KCL, the other capacitor current is

$$i_A = i_s + i_2 = \tfrac{1}{5} \text{ A} + \tfrac{1}{10} \text{ A} = \boxed{0.3 \text{ A}}$$

19. The homogeneous equation is

$$s^2 + 4s + 4 = 0$$

The roots are $(s + 2)^2$. The homogeneous solution is

$$Ae^{-2t} + Bte^{-2t}$$

The steady-state solution is

$$s \to j\omega$$

$$\omega = 4$$

$$\frac{d}{dt} \to j\omega$$

$$\frac{d^2}{dt^2} \to -\omega^2 = -16$$

The sine reference is used.

$$20 \text{ V} = (-16 \ \Omega + 16j \ \Omega + 4 \ \Omega)I_{\text{ss}}$$

$$I_{\text{ss}} = \frac{20}{-12 + j16} = -0.6 + j(-0.8) \text{ A}$$

$$i_{\text{ss}} = -0.6 \sin 4t - 0.8 \sin(4t + 90°) \text{ A}$$

$$= -0.6 \sin 4t - 0.8 \cos 4t \text{ A}$$

$$i = i_{\text{ss}} + Ae^{-2t} + Bte^{-2t}$$

$$i(0) = -0.8 \text{ A} + A = 0 \text{ A}$$

$$A = 0.8 \text{ A}$$

$$\frac{di}{dt} = -2.4 \cos 4t + 3.2 \sin 4t$$
$$- 2Ae^{-2t} + Be^{-2t} - 2Bte^{-2t}$$

$$\frac{di}{dt}(0) = -2.4 - 2A + B = 4 \text{ A}$$

$$B = 2.4 + 2A + 4 = 8 \text{ A}$$

$$i(t) = 0.8e^{-2t} + 8te^{-2t} - 0.6 \sin 4t - 0.8 \cos 4t \text{ A}$$

$$= \boxed{0.8e^{-2t} + 8te^{-2t} - \sin(4t + 53.1°) \text{ A}}$$

20. The initial charge on the 1 μF capacitor is

$$Q_0 = CV_0$$

$$= (10^{-6} \text{ F})(100 \text{ V})$$

$$= 10^{-4} \text{ C}$$

The initial energy, W, is

(a) $$W_0 = \tfrac{1}{2}C_1V_0^2$$

$$= \left(\frac{10^{-6} \text{ F}}{2}\right)(100 \text{ V})^2 = 0.005 \text{ J}$$

$$W_{1,\text{initial}} = 5 \text{ mJ}$$

In the final condition, the voltage across the two capacitors is the same.

$$W_{2,\text{initial}} = 0 \text{ mJ}$$

$$V_f = \frac{Q_1}{C_1} = \frac{Q_2}{C_2}$$

$$Q_2 = 2Q_1$$

As $Q_1 + Q_2 = Q_0$,

$$Q_1 = \tfrac{1}{3} \times 10^{-4} \text{ C}$$

$$Q_2 = \tfrac{2}{3} \times 10^{-4} \text{ C}$$

$$W_{\text{final}} = \tfrac{1}{2}\left(\frac{Q_1^2}{C_1}\right) + \tfrac{1}{2}\left(\frac{Q_2^2}{C_2}\right) \text{ J}$$

$$= \left(\tfrac{1}{2}\right)\left(10^6 \text{ F}^{-1}\right)\left(\tfrac{1}{9} \times 10^{-8} \text{ C}^2\right)$$

$$+ \left(\tfrac{1}{2}\right)\left(\frac{10^6 \text{ F}^{-1}}{2}\right)\left(\tfrac{4}{9} \times 10^{-8} \text{ C}^2\right)$$

$$W_{1,\text{final}} = 0.556 \text{ mJ}$$

$$W_{2,\text{final}} = 1.111 \text{ mJ}$$

$$W_{\text{loss}} = 3.333 \text{ mJ}$$

(b) $W_{\text{final}} = \tfrac{3}{18} \times 10^{-2} \text{ J} = \boxed{0.00167 \text{ J}}$

(c) $W_{\text{loss}} = 0.005 \text{ J} - 0.00167 \text{ J} = \boxed{0.00333 \text{ J}}$

31 Time Response

PRACTICE PROBLEMS

1. Which of the following is in the form of the solution to a first-order circuit?

(A) $\kappa + Ae^{\frac{-t}{\tau}}$

(B) $\kappa + Ae^{s_1 t} + Be^{s_2 t}$

(C) $\kappa + Ae^{st} + Bte^{st}$

(D) $\kappa + e^{\alpha t}\left(A\cos\beta t + B\sin\beta t\right)$

2. In the circuit shown, how long will it take for the capacitor voltage to reach 98% of its final value, assuming the capacitor is initially uncharged?

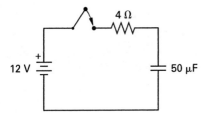

3. Determine the time-domain equation for the circuit shown.

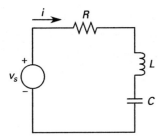

4. In the circuit shown in Prob. 3, the inductance is 80 mH and the capacitance is 20 μF. What is the minimum value of the resistance required for the circuit to be overdamped?

(A) 25 Ω

(B) 50 Ω

(C) 100 Ω

(D) 150 Ω

5. Complete the following table for a sine reference.

function	$A_p \angle \phi$	s
5		
$5e^{-200t}$		
$10\sin 377t$		
$10\sin\left(500t + 30°\right)$		
$20\cos\left(1000t + 60°\right)$		
$20e^{-200t}\sin 1000t$		

SOLUTIONS

1. The form of a first-order circuit solution is

$$x_t = \boxed{\kappa + Ae^{-t/\tau}}$$

The answer is (A).

2. The time constant, τ, is

$$\tau = RC = (4 \ \Omega)\left(50 \times 10^{-6} \ \text{F}\right)$$
$$= 200 \times 10^{-6} \ \text{s} \quad (200 \ \mu\text{s})$$

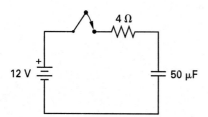

The build-up factor after a step change takes the form

$$1 - e^{-(t/\tau)}$$

Therefore, the time to 98% of the final value, regardless of the actual value, is

$$0.98 = 1 - e^{-(t/\tau)}$$

For the given circuit, $\tau = 200 \ \mu$s. Substituting and solving for the time gives

$$0.98 = 1 - e^{\frac{-t}{200 \times 10^{-6} \ \text{s}}}$$

$$-0.02 = -e^{-\frac{t}{200 \times 10^{-6} \ \text{s}}}$$

$$\ln 0.02 = -\frac{t}{200 \times 10^{-6} \ \text{s}}$$

$$t = -(\ln 0.02)\left(200 \times 10^{-6} \ \text{s}\right)$$

$$= \boxed{782.4 \times 10^{-6} \ \text{s} \quad (782 \ \mu\text{s} \quad [\approx 4\tau])}$$

3.

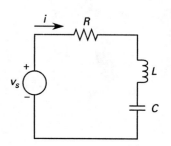

Using Kirchoff's voltage law (KVL) around the circuit gives

$$v_s - iR - v_L - v_C = 0 \ \text{V}$$

$$v_s = iR + v_L + v_C$$

$$= iR + L\left(\frac{di}{dt}\right) + \frac{1}{C}\int i \, dt$$

Differentiating gives

$$\frac{dv_s}{dt} = \left(\frac{di}{dt}\right)R + L\left(\frac{d^2i}{dt^2}\right) + \frac{i}{C}$$

$$= \boxed{L\left(\frac{d^2i}{dt^2}\right) + R\left(\frac{di}{dt}\right) + \frac{1}{C}i}$$

4. The overdamped condition is determined by the coefficients of the differential equation. Specifically, for the overdamped condition, $b^2 > 4ac$.

From the solution of Prob. 31.3,

$$R^2 > 4L\left(\frac{1}{C}\right)$$

$$R < \sqrt{(4)(80 \times 10^{-3} \ \text{H})\left(\frac{1}{20 \times 10^{-6} \ \text{F}}\right)} = \boxed{126.5 \ \Omega}$$

The answer is (D).

5.

function	$A_p \angle \phi$	s
5	$5\angle 0°$	$0 + j0$
$5e^{-200t}$	$5\angle 0°$	$-200 + j0$ Np/s
$10\sin 377t$	$10\angle 0°$	$0 \pm j377$ rad/s
$10\sin(500t + 30°)$	$10\angle 30°$	$0 \pm j500$ rad/s
$20\cos(1000t + 60°)$	$20\angle 150°$	$0 \pm j1000$ rad/s
$20e^{-200t}\sin 1000t$	$20\angle 0°$	$-200 \pm j1000$ s^{-1}

32 Frequency Response

PRACTICE PROBLEMS

1. For the circuit shown, determine the transfer function with $V(s)$ as the input and $I(s)$ as the output.

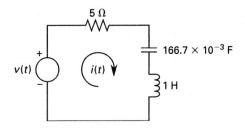

2. Plot the zeros and poles on the s domain for the circuit in Prob. 1.

3. Plot the magnitude Bode plot for the circuit in Prob. 1.

4. For the circuit shown, plot the idealized Bode plot of magnitude and phase.

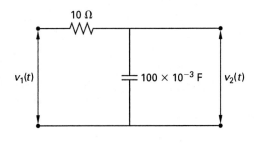

5. In the Bode plot for the circuit in Prob. 4, what type of filter is represented?
   (A) low-pass filter
   (B) high-pass filter
   (C) bandpass filter
   (D) notch filter

6. For the circuit shown, determine the (a) voltage transfer function, (b) frequency response, and (c) unit step voltage response.

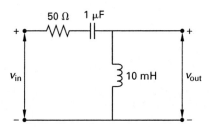

7. For the circuit shown, determine the (a) voltage transfer function, (b) frequency response, and (c) unit step voltage response.

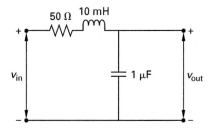

SOLUTIONS

1. The transfer function is

$$T_{net}(s) = \frac{I(s)}{V(s)} = \frac{1}{Z(s)}$$

First, transfer the circuit into the s domain.

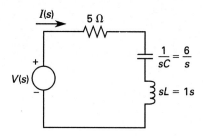

The impedance for this series circuit is found in terms of the complex frequency, s.

$$Z(s) = R + Z_C + Z_L = 5 + \frac{6}{s} + 1s$$

$$= \frac{5s + 6 + s^2}{s} = \frac{s^2 + 5s + 6}{s}$$

The transfer function is

$$\boxed{T_{net} = \frac{1}{Z(s)} = \frac{s}{s^2 + 5s + 6}}$$

2. Factor the transfer function into standard form.

$$T_{net}(s) = \frac{s}{s^2 + 5s + 6}$$

$$= \frac{s}{(s + 2)(s + 3)}$$

One zero at the origin is given by the numerator term. Two poles, one at -2 and the other at -3, exist. Plotting gives

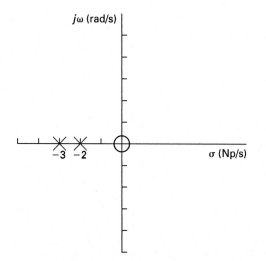

3. The transfer function was determined in Prob. 2 to be

$$T_{net}(s) = \frac{s}{(s + 2)(s + 3)}$$

The zero break frequency is $\omega = 1$. The break frequency for the poles is $\omega = 2$ and $\omega = 3$. These points are plotted on the following diagram.

For the zero, a $+20$ dB/decade line through the origin is plotted. For the poles, the low-frequency asymptotes are plotted on the frequency axis extending to lower frequencies from $\omega = 2$ and $\omega = 3$. The high-frequency asymptotes break down at -20 dB per decade from these points. Summing the asymptotic lines gives the net magnitude Bode plot as shown.

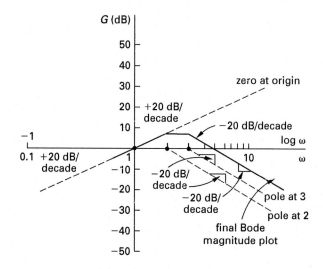

The value of the constant, K, is determined from

$$T_{net}(j\omega) = K\left(\frac{j\omega}{\left(1 + \frac{j\omega}{2}\right)\left(1 + \frac{j\omega}{3}\right)}\right)$$

$$K = A\left(\frac{\prod\limits_{1}^{n} z_n}{\prod\limits_{1}^{d} p_d}\right)$$

The transfer $T_{net}(s)$ gives the value of A as 1. For a zero at the origin in the s domain, the break frequency is ω equals 1. Therefore z_n equals 1. Thus,

$$K = (1)\left(\frac{1}{(2)(3)}\right) = 1/6$$

This makes $20 \log K$ equal to -15.563 dB. Either the Bode plot should be shifted downward by -15.563 or the zero dB point should be changed to -15.563 to obtain the final solution.

4. Because the calculations will take place in the s domain, transform the circuit.

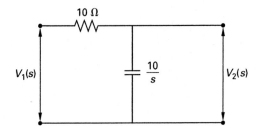

The transfer function, using the voltage divider concept, is

$$T_{\text{net}}(s) = \frac{V_2(s)}{V_1(s)} = \frac{\dfrac{10}{s}}{\dfrac{10}{s} + 10} = \frac{\dfrac{10}{s}}{\dfrac{10 + s10}{s}}$$

$$= \frac{10}{10 + s10}$$

$$= \frac{1}{s + 1}$$

The break frequency, or pole, occurs at $\omega = 1$. At this point the phase is $-45°$. Using the standard method, the resulting plot is

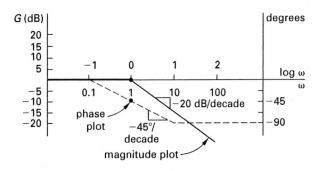

5. The Bode plot of Prob. 32.4 indicates that low frequencies are passed with a gain of 0 dB. At high frequencies, the output is attenuated and the gain steadily drops. The circuit is called a $\boxed{\text{low-pass filter.}}$

The answer is (A).

6. (a) By voltage division,

$$\left(\frac{V_{\text{out}}}{V_{\text{in}}}\right)(s) = \frac{0.01s}{0.01s + 50 + \dfrac{10^6}{s}}$$

$$= \boxed{\frac{s^2}{s^2 + 5000s + 10^8}}$$

(b) The denominator needs an underdamped response, so it is put into standard form to take advantage of the Laplace transform pairs table.

$$\left(\frac{V_{\text{out}}}{V_{\text{in}}}\right)(s) = \frac{s^2}{(s + 2500)^2 + (9682)^2}$$

For the frequency response $s \to j\omega$,

$$\left(\frac{V_{\text{out}}}{V_{\text{in}}}\right)(j\omega) = \frac{-\omega^2}{10^8 - \omega^2 + j5000\omega}$$

At low frequency,

$$\frac{V_{\text{out}}}{V_{\text{in}}} \to \frac{-\omega^2}{10^8}$$

This has a slope of 40 dB per decade and a phase of $180°$ on a Bode diagram.

At high frequency,

$$\frac{V_{\text{out}}}{V_{\text{in}}} \to 1$$

At $\omega = 10^4$ rad/s,

$$\frac{V_{\text{out}}}{V_{\text{in}}} = 2\angle 90°$$

The frequency response value is $\boxed{6 \text{ dB.}}$

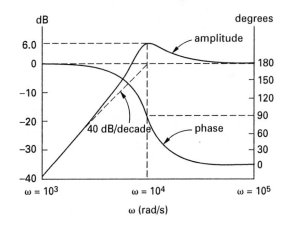

(c) The step response with $V_{\text{in}} = 1/s$ is

$$V_{\text{out}} = \frac{s}{(s + 2500)^2 + (9682)^2} \text{ V}$$

$$= \frac{s + 2500}{(s + 2500)^2 + (9682)^2} \text{ V}$$

$$- \left(\frac{2500}{9682}\right)\left(\frac{9682}{(s + 2500)^2 + (9682)^2}\right) \text{ V}$$

$$v_{\text{out}}(t) = e^{-2500t}(\cos 9682t - 0.258 \sin 9682t) \text{ V}$$

$$= \boxed{1.03e^{-2500t} \cos(9682t + 14.5°) \text{ V}}$$

7. (a) By voltage division,

$$\frac{V_{\text{out}}}{V_{\text{in}}} = \frac{\dfrac{10^6}{s}}{10^{-2}s + 50 + \dfrac{10^6}{s}}$$

$$= \boxed{\frac{10^8}{s^2 + 5000s + 10^8}}$$

$$\left(\frac{V_{\text{out}}}{V_{\text{in}}}\right)(j\omega) = \frac{10^8}{10^8 - \omega^2 + j5000\omega}$$

(b) At low frequency,

$$\frac{V_{\text{out}}}{V_{\text{in}}} \to 1\angle 0°$$

At high frequency,

$$\frac{V_{\text{out}}}{V_{\text{in}}} \to \frac{10^8}{-\omega^2}$$

On the Bode diagram, this has a slope of -40 dB per decade and an angle of $-180°$. At $\omega = 10^4$ rad/s,

$$\frac{V_{\text{out}}}{V_{\text{in}}} = 2\angle -90°$$

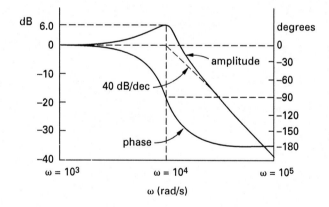

(c) The step response with $V_{\text{in}} = 1/s$ is

$$V_{\text{out}} = \frac{10^8}{s(s^2 + 5000s + 10^8)} \text{ V·s}$$

This is expanded in partial fractions to

$$V_{\text{out}} = \frac{1}{s} - \frac{s + 5000}{(s + 2500)^2 + (9682)^2}$$

This is recast to fit the Laplace transform pairs table.

$$V_{\text{out}} = \frac{1}{s} - \frac{s + 2500}{(s + 2500)^2 + (9682)^2}$$

$$- \frac{\left(\dfrac{2500}{9682}\right)(9682)}{(s + 2500)^2 + (9682)^2}$$

$$v_{\text{out}}(t) = 1 - e^{-2500t}(\cos 9682t + 0.258\sin 9682t) \text{ V}$$

$$= \boxed{1 - 1.033e^{-2500t}\cos(9682t - 14.5°) \text{ V}}$$

4. Because the calculations will take place in the s domain, transform the circuit.

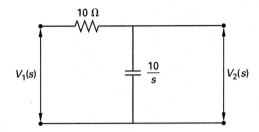

The transfer function, using the voltage divider concept, is

$$T_{\text{net}}(s) = \frac{V_2(s)}{V_1(s)} = \frac{\dfrac{10}{s}}{\dfrac{10}{s} + 10} = \frac{\dfrac{10}{s}}{\dfrac{10 + s10}{s}}$$

$$= \frac{10}{10 + s10}$$

$$= \frac{1}{s + 1}$$

The break frequency, or pole, occurs at $\omega = 1$. At this point the phase is $-45°$. Using the standard method, the resulting plot is

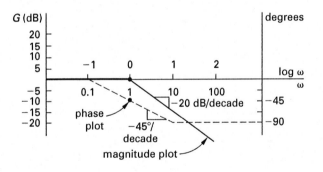

5. The Bode plot of Prob. 32.4 indicates that low frequencies are passed with a gain of 0 dB. At high frequencies, the output is attenuated and the gain steadily drops. The circuit is called a $\boxed{\text{low-pass filter.}}$

The answer is (A).

6. (a) By voltage division,

$$\left(\frac{V_{\text{out}}}{V_{\text{in}}}\right)(s) = \frac{0.01s}{0.01s + 50 + \dfrac{10^6}{s}}$$

$$= \boxed{\frac{s^2}{s^2 + 5000s + 10^8}}$$

(b) The denominator needs an underdamped response, so it is put into standard form to take advantage of the Laplace transform pairs table.

$$\left(\frac{V_{\text{out}}}{V_{\text{in}}}\right)(s) = \frac{s^2}{(s + 2500)^2 + (9682)^2}$$

For the frequency response $s \to j\omega$,

$$\left(\frac{V_{\text{out}}}{V_{\text{in}}}\right)(j\omega) = \frac{-\omega^2}{10^8 - \omega^2 + j5000\omega}$$

At low frequency,

$$\frac{V_{\text{out}}}{V_{\text{in}}} \to \frac{-\omega^2}{10^8}$$

This has a slope of 40 dB per decade and a phase of $180°$ on a Bode diagram.

At high frequency,

$$\frac{V_{\text{out}}}{V_{\text{in}}} \to 1$$

At $\omega = 10^4$ rad/s,

$$\frac{V_{\text{out}}}{V_{\text{in}}} = 2\angle 90°$$

The frequency response value is $\boxed{6 \text{ dB.}}$

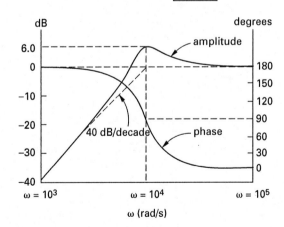

(c) The step response with $V_{\text{in}} = 1/s$ is

$$V_{\text{out}} = \frac{s}{(s + 2500)^2 + (9682)^2} \text{ V}$$

$$= \frac{s + 2500}{(s + 2500)^2 + (9682)^2} \text{ V}$$

$$- \left(\frac{2500}{9682}\right)\left(\frac{9682}{(s + 2500)^2 + (9682)^2}\right) \text{ V}$$

$$v_{\text{out}}(t) = e^{-2500t}\left(\cos 9682t - 0.258\sin 9682t\right) \text{ V}$$

$$= \boxed{1.03e^{-2500t}\cos(9682t + 14.5°) \text{ V}}$$

7. (a) By voltage division,

$$\frac{V_{\text{out}}}{V_{\text{in}}} = \frac{\dfrac{10^6}{s}}{10^{-2}s + 50 + \dfrac{10^6}{s}}$$

$$= \boxed{\frac{10^8}{s^2 + 5000s + 10^8}}$$

$$\left(\frac{V_{\text{out}}}{V_{\text{in}}}\right)(j\omega) = \frac{10^8}{10^8 - \omega^2 + j5000\omega}$$

(b) At low frequency,

$$\frac{V_{\text{out}}}{V_{\text{in}}} \rightarrow 1\angle 0^\circ$$

At high frequency,

$$\frac{V_{\text{out}}}{V_{\text{in}}} \rightarrow \frac{10^8}{-\omega^2}$$

On the Bode diagram, this has a slope of -40 dB per decade and an angle of -180°. At $\omega = 10^4$ rad/s,

$$\frac{V_{\text{out}}}{V_{\text{in}}} = 2\angle -90^\circ$$

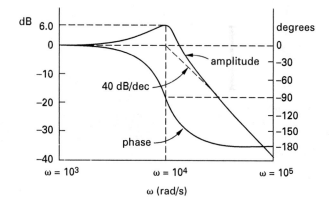

(c) The step response with $V_{\text{in}} = 1/s$ is

$$V_{\text{out}} = \frac{10^8}{s(s^2 + 5000s + 10^8)} \text{ V·s}$$

This is expanded in partial fractions to

$$V_{\text{out}} = \frac{1}{s} - \frac{s + 5000}{(s + 2500)^2 + (9682)^2}$$

This is recast to fit the Laplace transform pairs table.

$$V_{\text{out}} = \frac{1}{s} - \frac{s + 2500}{(s + 2500)^2 + (9682)^2}$$

$$- \frac{\left(\dfrac{2500}{9682}\right)(9682)}{(s + 2500)^2 + (9682)^2}$$

$$v_{\text{out}}(t) = 1 - e^{-2500t}(\cos 9682t + 0.258 \sin 9682t) \text{ V}$$

$$= \boxed{1 - 1.033e^{-2500t}\cos(9682t - 14.5^\circ) \text{ V}}$$

33 Generation Systems

PRACTICE PROBLEMS

1. What term describes the process of minimizing irreversibilities by raising the temperature of the water returning to a steam generator?
- (A) superheat
- (B) reheat
- (C) regeneration
- (D) both (A) and (C)

2. What fission rate (fissions/s) is required for a nuclear plant to have a 1000 MW thermal output?

3. What type of first stage, or control stage, is normally used on large turbines used for electric power generation?
- (A) impulse
- (B) Rateau
- (C) Curtis
- (D) Parsons

4. In the load sharing diagram shown, the speed droop for both generators is 0.6 Hz/1000 kW. If generator A is to carry all the load, what no-load frequency setpoint is required on generator A while maintaining the system frequency at 60 Hz?

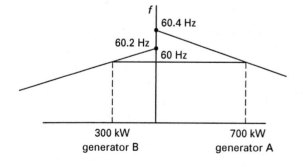

5. Draw the load sharing diagram for Prob. 4 with generator A carrying all the load. Label all significant points.

6. The field of a DC generator is normally located on the
- (A) stator
- (B) rotor
- (C) armature
- (D) round rotor

7. What is the maximum speed for 60 Hz AC generator operation?

8. What power quality term describes a transient of short duration and high magnitude?
- (A) overvoltage
- (B) harmonic
- (C) sag
- (D) surge

SOLUTIONS

1. Superheat raises the temperature of the steam supplied to the turbine. Reheat raises the temperature of the steam at the input of the low-temperature stages of a turbine. Regeneration raises the temperature of the water returning to the steam generator.

The answer is (C).

2. The fission rate, fr, is in units of fissions/s. The average energy released per fission is 200 MeV. Unit analysis gives the power output, P_s,

$$P = \left(\frac{\text{fissions}}{\text{s}}\right)\left(\frac{\text{energy}}{\text{fission}}\right)$$

$$= \text{fr}\left(200 \; \frac{\text{MeV}}{\text{fission}}\right)$$

$$\text{fr} = \frac{P}{200 \; \dfrac{\text{MeV}}{\text{fission}}}$$

$$= \frac{1000 \times 10^6 \text{ W}}{\left(200 \times 10^6 \; \dfrac{\text{eV}}{\text{fission}}\right)\left(\dfrac{1.602 \times 10^{-19}\text{J}}{1 \text{ eV}}\right)}$$

$$= \boxed{3.12 \times 10^{19} \text{ fissions/s}}$$

3. Large turbines use impulse control stages because no pressure drop occurs across such a stage, thereby allowing partial arc admission of the steam without significant forces being exerted on the turbine rotor.

The answer is (A).

4.

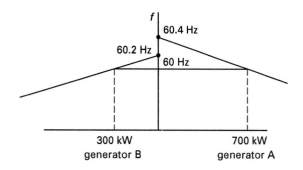

300 kW
generator B

700 kW
generator A

The power carried by an AC generator is

$$P = \frac{f_{\text{nl}} - f_{\text{sys}}}{f_{\text{droop}}}$$

Solving for the no-load frequency gives

$$f_{\text{nl}} = P f_{\text{droop}} + f_{\text{sys}}$$

Solving for the no-load frequency increase necessary to carry an additional 300 kW gives

$$\Delta f_{\text{nl}} = \Delta P f_{\text{droop}}$$

$$= (300 \text{ kW})\left(\frac{0.6 \text{ Hz}}{1000 \text{ kW}}\right)$$

$$= 0.18 \text{ Hz}$$

The required no-load frequency is

$$f_{\text{nl}} = f_{\text{nl,current}} + \Delta f_{\text{nl}}$$

$$= 60.4 \text{ Hz} + 0.18 \text{ Hz}$$

$$= \boxed{60.58 \text{ Hz} \quad (60.6 \text{ Hz})}$$

Generator B's frequency will have to be lowered by 0.2 Hz to maintain the system frequency at 60 Hz.

5. The real load sharing diagram with generator A carrying all the load is

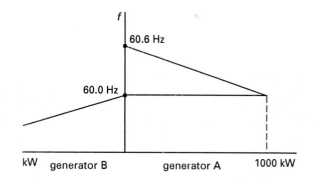

6. The field of a DC machine is normally on the stationary portion, or stator. This allows commutation to be accomplished on the rotor.

The answer is (A).

7. Synchronous speed is derived by

$$n_s = \frac{120f}{P}$$

The minimum number of poles determines the maximum speed. Two is the minimum number. Thus,

$$n_s = \frac{(120)(60 \text{ Hz})}{2} = \boxed{3600 \text{ rpm}}$$

8. A transient condition of short duration and high magnitude is called a surge.

The answer is (D).

34 Three-Phase Electricity and Power

PRACTICE PROBLEMS

1. A three-phase 208 V (rms) system supplies heating elements connected in a wye configuration. What is the resistance of each element if the total balanced load is 3 kW?

2. A three-phase system has a balanced delta load of 5000 kW at 84% power factor. If the line voltage is 4160 V (rms), what is the line current?

3. A 120 V (per phase, rms) three-phase system has a balanced load consisting of three 10 Ω resistances. What total power is dissipated if the connection is (a) a wye configuration and (b) a delta configuration?

4. A balanced delta load consists of three 20 Ω∠25° impedances. The 60 Hz line voltage is 208 V (rms). Find the (a) phase current, (b) line current, (c) phase voltage, (d) power consumed by each phase, and (e) total power.

5. A balanced wye load consists of three $3 + j4\,\Omega$ impedances. The system voltage is 110 V (rms). Find the (a) phase voltage, (b) line current, (c) phase current, and (d) total power.

6. Consider the one-line diagram of a three-phase distribution system. What is the generator base voltage?

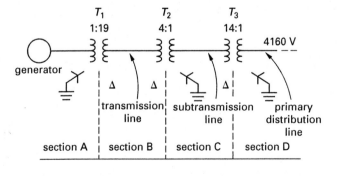

SOLUTIONS

1.
$$R = \frac{V_p^2}{P_p} = \frac{\left(\dfrac{V_l}{\sqrt{3}}\right)^2}{\dfrac{P_t}{3}} = \frac{V_l^2}{P}$$
$$= \frac{(208 \text{ V})^2}{3 \times 10^3 \text{ W}}$$
$$= \boxed{14.42\ \Omega}$$

2.
$$I_l = \left(\frac{1}{\sqrt{3}}\right)\left(\frac{P_\Delta}{V_l \cos\phi}\right)$$
$$= \left(\frac{1}{\sqrt{3}}\right)\left(\frac{5000 \times 10^3 \text{ W}}{(4160 \text{ V})(0.84)}\right)$$
$$= \boxed{826 \text{ A}}$$

3. (a) $P_p = \dfrac{V_p^2}{R} = \dfrac{(120 \text{ V})^2}{10\ \Omega} = 1440 \text{ W}$

$P_t = 3P_p = (3)(1440 \text{ W}) = \boxed{4320 \text{ W}}$

(b) $P_t = 3\left(\dfrac{V_l^2}{R}\right) = (3)\left(\dfrac{(120 \text{ V})^2}{10\ \Omega}\right)$

$= \boxed{4320 \text{ W}}$

4. (a) $\mathbf{I}_p = \dfrac{\mathbf{V}_l}{\mathbf{Z}} = \dfrac{208 \text{ V}\angle 0°}{20\ \Omega\angle 25°}$

$= \boxed{10.4 \text{ A}\angle{-25°}}$

(b) $\mathbf{I}_l = \sqrt{3}\ \angle{-30°}\ \mathbf{I}_p$

$= \left(\sqrt{3}\ \angle{-30°}\right)(10.4 \text{ A}\angle{-25°})$

$= \boxed{18 \text{ A}\angle{-55°}}$

(c) $\mathbf{V}_p = \mathbf{V}_l$

$= \boxed{208 \text{ V}\angle 0°}$

Power—Generation

(d)
$$P_p = V_p I_p \cos \phi$$
$$= (208 \text{ V})(10.4 \text{ A}) \cos 25°$$
$$= \boxed{1960.5 \text{ W}}$$

(e)
$$P_\Delta = 3P_p = (3)(1960.5 \text{ W})$$
$$= \boxed{5882 \text{ W}}$$

5. (a)
$$V_p = \frac{110 \text{ V}}{\sqrt{3}} = \boxed{63.51 \text{ V}}$$

The impedance is given as $3+j4 \ \Omega$ in rectangular coordinates. The magnitude of the impedance, Z, in complex algebra, is

(b)
$$Z = \sqrt{(3 \ \Omega)^2 + (4 \ \Omega)^2} = 5 \ \Omega$$
$$\phi = \arctan \frac{4 \ \Omega}{3 \ \Omega} = 53.15°$$
$$\mathbf{I}_l = \frac{\mathbf{V}_p}{\mathbf{Z}} = \frac{63.51 \text{ V} \angle 0°}{5 \ \Omega \angle 53.13°}$$
$$= \boxed{12.70 \text{ A} \angle -53.13°}$$

(c)
$$\mathbf{I}_p = \mathbf{I}_l = \boxed{12.70 \text{ A} \angle -53.13°}$$

(d)
$$P_t = \sqrt{3} V I \cos \phi$$
$$= \sqrt{3}(110 \text{ V})(12.7 \text{ A})(\cos 53.13°)$$
$$= \boxed{1451 \text{ W}}$$

6. For a three-phase system, the line voltage is the base. The base in the primary distribution line is given as 4160 V, or $V = 1.0$ pu $\angle 0°$.

Because the transformers scale the base values in each section, the base voltage value of 4160 V will be determined at the generator. The $\sqrt{3}$ is used to account for the wye connection line-to-phase and phase-to-line conversions.

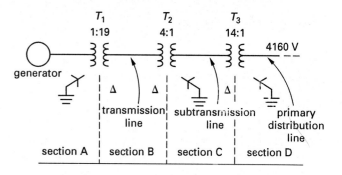

$$V_{\text{gen,base}} = V_{\text{gen,line}}$$
$$= \left(\frac{V_{\text{line,D}}}{\sqrt{3}}\right)\left(\frac{N_{3,\text{pri}}}{N_{3,\text{sec}}}\right)\left(\frac{1}{\sqrt{3}}\right)$$
$$\times \left(\frac{N_{2,\text{pri}}}{N_{2,\text{sec}}}\right)\left(\frac{N_{1,\text{pri}}}{N_{1,\text{sec}}}\right)\sqrt{3}$$
$$= \left(\frac{4160 \text{ V}}{\sqrt{3}}\right)\left(\frac{14}{1}\right)\left(\frac{1}{\sqrt{3}}\right)\left(\frac{4}{1}\right)\left(\frac{1}{19}\right)\sqrt{3}$$
$$= \boxed{7070 \text{ V} \quad (7.07 \text{ kV})}$$

This generator line voltage of 7.07 kV is the 1.0 pu $\angle 0°$ value in section A of the distribution system.

35 Power Distribution

PRACTICE PROBLEMS

1. Which of the following voltages is NOT commonly used in the subtransmission portion of the power distribution system?
  (A) 25 kV
  (B) 69 kV
  (C) 115 kV
  (D) 138 kV

2. A single-conductor cable with an outside diameter of 0.5 cm is to be used in an underground installation. What is the optimal conductor radius?
  (A) 0.05 cm
  (B) 0.1 cm
  (C) 0.2 cm
  (D) 0.3 cm

3. A single-conductor capacitance-graded cable with an outside diameter of 10 cm can withstand a maximum electric field of 700 kV/m before insulation breakdown occurs. What is the maximum operating voltage of the conductor?

4. What is the total capacitance of 100 m of single-conductor cable with the optimal radii ratio and polyethylene insulation ($\epsilon_r = 2.25$)?

5. A three-phase 11.5 kV generator drives a 500 kW, 0.866 lagging power factor load. Determine the (a) line current and (b) necessary generator rating.

6. A transmission line with 8% reactance on a 200 MVA base connects two substations. At the first substation the voltage is 1.03 pu∠5°, and at the second the voltage is 0.98 pu∠−2.5°. Find the (a) power and (b) volt-amperes reactive flowing out of the first substation.

SOLUTIONS

1. The trend in transmission voltages is toward higher voltages. The most common are 115 kV, 69 kV, 138 kV, and 238 kV. Thus, 25 kV is not common.

The answer is (A).

2. The optimal ratio is

$$\frac{r_2}{r_1} = 2.718$$

The outside diameter is given as 0.5 cm. Thus, the radius, r_2, is 0.25 cm. Substituting gives

$$r_1 = \frac{r_2}{2.718} = \frac{0.25 \text{ cm}}{2.718} = \boxed{0.0920 \text{ cm} \quad (0.1 \text{ cm})}$$

The answer is (B).

3. The operating voltage for a capacitance-graded cable is

$$V = E_{\max}\left(r_1 \ln \frac{r_2}{r_1} + r_2 \ln \frac{r_3}{r_2}\right)$$
$$= E_{\max}\left(r_1 \ln \xi + r_2 \ln \xi\right)$$

The values of r_1 and r_2 are the only unknowns. They are determined from the ratio of radii, ξ.

$$\xi = \frac{r_3}{r_2}$$

$$r_2 = \frac{r_3}{\xi} = \frac{\left(\dfrac{10 \text{ cm}}{2}\right)}{2.718} = 1.84 \text{ cm}$$

$$\xi = \frac{r_2}{r_1}$$

$$r_1 = \frac{r_2}{\xi} = \frac{1.84 \text{ cm}}{2.718} = 0.677 \text{ cm}$$

Substituting known and calculated values gives

$$V = E_{\max}\left(r_1 \ln \xi + r_2 \ln \xi\right)$$
$$= \left(700 \; \frac{\text{kV}}{\text{m}}\right)\left(\begin{array}{c}(0.677 \text{ cm})\ln 2.718 \\ + (1.84 \text{ cm})\ln 2.718\end{array}\right)\left(\frac{1 \text{ m}}{100 \text{ cm}}\right)$$
$$= \boxed{17.6 \text{ kV}}$$

4. The capacitance per unit length is

$$C_l = \frac{Q}{V} = \frac{2\pi\epsilon}{\ln\xi}$$

$$= \frac{2\pi\epsilon_r\epsilon_0}{\ln e}$$

$$= \frac{2\pi\epsilon_r\epsilon_0}{1}$$

$$= 2\pi\,(2.25)\left(8.854\times10^{-12}\,\frac{\text{F}}{\text{m}}\right)$$

$$= 125\times10^{-12}\,\text{F/m}$$

The total capacitance is

$$C = C_l L$$

$$= \left(125\times10^{-12}\,\frac{\text{F}}{\text{m}}\right)(100\,\text{m})$$

$$= \boxed{125\times10^{-10}\,\text{F}}$$

5. (a)

$$P_{\text{phase}} = I_p V_p(\text{pf})$$

$$I_p = \frac{P_{\text{phase}}}{V_p(\text{pf})}$$

$$I_l = \left(\frac{P_p}{V_l(\text{pf})}\right)\sqrt3$$

$$= \left(\frac{\frac{P_{\text{total}}}{3\text{ phases}}}{V_l}\right)\sqrt3$$

$$= \frac{\frac{500\text{ kW}}{3\text{ phases}}}{\left(\frac{11.5\text{ kV}}{\sqrt3}\right)(0.866)}$$

$$= \boxed{29\text{ A}}$$

(b)

$$P = S(\text{pf})$$

$$S = \frac{P}{\text{pf}}$$

$$= \frac{500\text{ kW}}{0.866}$$

$$= \boxed{577\text{ kVA}}$$

6.

(a)

$$I_{\text{pu}} = \frac{V_{\text{pu}}}{Z_{\text{pu}}} = \frac{1.03\text{ V}\angle5° - 0.98\text{ V}\angle-2.5°}{j0.08}$$

$$= 1.758\text{ pu}\angle-19.5°$$

$$S_{\text{pu}} = V_{\text{pu}}I_{\text{pu}}^*$$

$$= (1.03\text{ pu}\angle5°)(1.758\text{ pu}\angle19.5°)$$

$$= 1.811\text{ pu}\angle24.5°$$

$$= 1.648 + j0.751\text{ pu}$$

$$P = P_{\text{pu}}P_{\text{base}} = (1.648)(200\text{ MW})$$

$$= \boxed{330\text{ MW}}$$

(b)

$$Q = Q_{\text{pu}}Q_{\text{base}} = (0.751)(200\text{ MVAR})$$

$$= \boxed{150\text{ MVAR}}$$

36 Power Transformers

PRACTICE PROBLEMS

1. Which electrical element in the exact (real) transformer model accounts for the eddy current and hysteresis losses?

(A) B_c

(B) R_c

(C) R_p

(D) R_s

2. Which electrical element in the exact (real) transformer model accounts for magnetic flux leakage?

(A) B_c

(B) X_p

(C) X_s

(D) (B) or (C)

3. What is the eddy current power loss for a 500 kVA rated, 200 kg iron-core transformer with a coupling coefficient of 4×10^{-4} in a 1.4 T peak magnetic field?

(A) 3.0 W

(B) 9.0 W

(C) 400 W

(D) 570 W

4. A transformer is rated for 34.5 kV. The measured secondary voltage at no load is 35.02 kV. What is the percent voltage regulation?

(A) 0.50%

(B) 1.4%

(C) 1.5%

(D) 2.0%

5. An open-circuit test is conducted on a 120 V transformer winding. The results are $V_{1oc} = 120$ V, $V_{2oc} = 240$ V, $I_{1oc} = 0.25$ A, and $P_{oc} = 20$ W. What apparent power is necessary to supply core losses in this transformer?

(A) 20 VA

(B) 30 VA

(C) 60 VA

(D) 120 VA

6. A 13.8 kV single-phase transformer is subjected to an open-circuit test with the following results: $P_{in} = 900$ W, $I_{in} = 0.2$ A, and $V_{out} = 460$ V (secondary). Determine the values of the equivalent circuit parameters that can be found from this test.

7. A 32 kVA single-phase transformer is rated at 1.5 kV on the primary. A 60 Hz short-circuit test on the primary indicates $P_{in} = 1$ kW, $I_{in} = 20$ A, $I_s = 100$ A, and $V_p = 80$ V. Determine the (a) winding resistances and (b) leakage inductances.

8. A single-phase transformer is rated at 115 kV at 500 kVA. A short-circuit test on the high-voltage side at rated current indicates $P_{sc} = 435$ W and $V_{sc} = 2.5$ kV. Determine the per-unit (pu) impedance of the transformer.

SOLUTIONS

1. Core losses include those caused by eddy currents and hysteresis. In the exact transformer model, these are accounted for by $G_c = 1/R_c$.

The answer is (B).

2. Magnetic flux leakage is a term used to describe any flux that is not mutual. The primary and secondary self-inductances represented by X_p and X_s are not mutual.

The answer is (D).

3. The eddy current loss is found from

$$P_e = k_e B_m^2 f^2 m$$

Assume 60 Hz is the frequency. Substitute the given values.

$$P_e = (4 \times 10^{-4})(1.4 \text{ T})^2 (60 \text{ Hz})^2 (200 \text{ kg})$$

$$= \boxed{565 \text{ W} \quad (570 \text{ W})}$$

The answer is (D).

4. The voltage regulation is

$$\text{VR} = \frac{V_{\text{nl}} - V_{\text{fl}}}{V_{\text{fl}}}$$

$$= \frac{\dfrac{V_p}{a} - V_{s,\text{rated}}}{V_{s,\text{rated}}}$$

Because the voltage was measured at the output of the secondary, it is equivalent to V_p/a. Thus, the voltage regulation is

$$\text{VR} = \frac{35.02 \text{ V} - 34.5 \text{ V}}{34.5 \text{ V}} = 15.07 \times 10^{-3}$$

To express as a percentage, multiply by 100.

$$\text{VR} = \boxed{1.5\%}$$

The answer is (C).

5. The open-circuit test determines the parameters associated with core losses. The apparent power is

$$S_{\text{oc}} = I_{\text{1oc}} V_{\text{1oc}}$$

$$= (0.25 \text{ A})(120 \text{ V})$$

$$= \boxed{30 \text{ VA}}$$

The answer is (B).

6. The open-circuit (oc) test obtains the core parameters G and B.

$$G = \frac{P_{\text{oc}}}{V_{\text{oc}}^2} = \frac{900 \text{ W}}{(13{,}800 \text{ V})^2} = \boxed{4.73 \times 10^{-6} \text{ S}}$$

$$Q^2 = S^2 - P^2 = \left((13{,}800 \text{ VA})(0.2)\right)^2 - (900 \text{ W})^2$$

$$Q = 2609 \text{ VAR}$$

$$B = \frac{Q}{V^2} = \frac{2609 \text{ VAR}}{(13{,}800 \text{ V})^2} = \boxed{13.7 \times 10^{-6} \text{ S}}$$

The turns ratio is

$$\frac{V_p}{N_p} = \frac{V_s}{N_s}$$

$$\frac{N_p}{N_s} = \frac{V_p}{V_s} = \frac{13{,}800 \text{ V}}{460 \text{ V}}$$

$$= \boxed{30}$$

7. (a)
$$N_p I_p = N_s I_s$$

$$a = \frac{N_p}{N_s} = \frac{I_s}{I_p}$$

$$= \frac{100 \text{ A}}{20 \text{ A}}$$

$$= 5$$

$$P_{\text{sc}} = I_p^2 R_p + I_s^2 R_s = 1000 \text{ W}$$

$$I_p^2 (R_p + a^2 R_s) = 1000 \text{ W}$$

$$R_p + a^2 R_s = \frac{1000 \text{ W}}{(20 \text{ A})^2} = 2.5 \text{ }\Omega$$

Assume $R_p = a^2 R_s = \boxed{1.25 \text{ }\Omega}$

$$R_s = \frac{1.25 \text{ }\Omega}{(5)^2} = \boxed{0.05 \text{ }\Omega}$$

(b)
$$S = V_p I_p = (80 \text{ V})(20 \text{ A}) = 1600 \text{ VA}$$
$$= P + jQ$$
$$Q^2 = S^2 - P^2$$
$$Q = 1249 \text{ VAR}$$
$$= I_1^2(X_p + a^2 X_s)$$
$$X_p + a^2 X_s = \frac{Q}{I_1^2} = \frac{1249 \text{ VAR}}{(20 \text{ A})^2} = 3.12 \ \Omega$$

Assume $X_p = a^2 X_s$, then

$$2X_p = 3.12$$
$$X_p = \frac{3.12}{2} = 1.56 \ \Omega = \omega L_p = 377 L_p$$
$$L_p = \frac{1.56 \ \Omega}{377 \ \frac{\text{rad}}{\text{s}}} = \boxed{4.14 \text{ mH}}$$
$$X_s = \frac{X_p}{a^2} = \frac{1.56}{a^2} = 0.0624 \ \Omega = \omega L_s = 377 L_s$$
$$L_s = \frac{0.0624 \ \Omega}{377 \ \frac{\text{rad}}{\text{s}}} = \boxed{0.166 \text{ mH}}$$

8. The per-unit impedance is found from

$$Z_{\text{pu}} = \frac{Z_{\text{actual}}}{Z_{\text{base}}} = \frac{R + jX}{Z_{\text{base}}} \qquad \text{[I]}$$

Find the unknown quantities.

$$Z_{\text{base}} = \frac{V_{\text{base}}}{I_{\text{base}}} \qquad \text{[II]}$$

The short-circuit test is performed at rated current, which is I_{base}. Thus,

$$I_{\text{rated}} = I_{\text{base}} = \frac{S_{\text{rated}}}{V_{\text{rated}}}$$
$$= \frac{500 \text{ kVA}}{115 \text{ kV}}$$
$$= 4.35 \text{ A}$$

Substitute the rated current into Eq. II.

$$Z_{\text{base}} = \frac{115 \text{ kV}}{4.35 \text{ A}} = 26{,}440 \ \Omega$$

Find R from the short circuit (sc) parameters.

$$P_{\text{sc}} = I_{\text{sc}}^2 R = I_{\text{rated}}^2 R$$
$$R = \frac{P_{\text{sc}}}{I_{\text{rated}}^2} = \frac{435 \text{ W}}{(4.35 \text{ A})^2} = 23 \ \Omega$$

Find X from

$$Q_{\text{sc}} = I_{\text{rated}}^2 X$$
$$X = \frac{Q}{I_{\text{rated}}^2} \qquad \text{[III]}$$

The reactive power is found using

$$S_{\text{sc}}^2 = P_{\text{sc}}^2 + Q_{\text{sc}}^2$$
$$Q_{\text{sc}} = \sqrt{S_{\text{sc}}^2 - P_{\text{sc}}^2}$$
$$= \sqrt{(I_{\text{rated}} V_{\text{sc}})^2 - (435 \text{ W})^2}$$
$$= \sqrt{\left((4.35 \text{ A})(2.5 \times 10^3 \text{ V})\right)^2 - (435 \text{ W})^2}$$
$$= 10{,}870 \text{ VAR}$$

Substitute into Eq. III, giving

$$X = \frac{10{,}870 \text{ VAR}}{(4.35 \text{ A})^2} = 574 \ \Omega$$

Substitute the calculated values for R, X, and Z_{base} into Eq. I.

$$Z_{\text{pu}} = \frac{23 \ \Omega + j574 \ \Omega}{26{,}440 \ \Omega} = 0.0009 + j0.022 \text{ pu}$$
$$= \boxed{0.001 + j0.022 \text{ pu}}$$

37 Power Transmission Lines

PRACTICE PROBLEMS

1. As frequency increases, the skin effect causes which of the following parameters to increase?

(A) AC resistance

(B) DC resistance

(C) internal inductance

(D) external inductance

2. What distributed parameter is normally insignificant and ignored in calculations?

(A) line resistance

(B) line inductance

(C) shunt resistance

(D) shunt capacitance

3. The velocity factor is defined as the ratio of the velocity of propagation, v_w, to the speed of light, c. What is the velocity factor in a single waxwing conductor at 60 Hz?

(A) 0.29

(B) 0.61

(C) 0.90

(D) 0.97

4. What is parameter A of a 220 km falcon line with $Z_l = 0.85 \ \Omega/\text{mi} \angle 70°$ and $Y_l = 7 \times 10^{-6} \ \text{S/mi} \angle 90°$?

(A) $4.0 \times 10^{-4} \angle 160°$

(B) $0.948 \angle 1.2°$

(C) $0.996 \angle 0.008°$

(D) $1.0 \angle 20°$

5. A three-phase, 215 kV transmission line supplies a 100 MW wye-connected load at 0.8 pf lagging. The ABCD parameters are A = $0.9 \angle 2.0°$, B = $150 \angle 80°$, C = $0.001 \angle 90°$, and D = $0.9 \angle 2°$. What is the sending-end voltage?

SOLUTIONS

1. The skin effect causes AC resistance to increase and internal inductance to decrease.

The answer is (A).

2. The shunt resistance is normally very large and thus has a minimal effect on current and voltage.

The answer is (C).

3. The velocity of wave propagation is

$$v_w = \frac{1}{\sqrt{L_l C_l}}$$

From App. 37.A, a waxwing conductor has an inductive reactance of

$$X_{L,l} = 0.476 \ \Omega/\text{mi}$$

Solving for the inductance per unit length gives

$$X_{L,l} = \omega L = 2\pi f L$$

$$L = \frac{X_{L,l}}{2\pi f} = \frac{0.476 \ \dfrac{\Omega}{\text{mi}}}{2\pi \ (60 \ \text{s}^{-1})}$$

$$= 1.26 \times 10^{-3} \ \text{H/mi}$$

The capacitive reactance, from App. 37.A, is

$$X_C = 0.1090 \ \text{M}\Omega\text{-mi}$$

Solving for C_l gives

$$|X_C| = \frac{1}{\omega C_l} = \frac{1}{2\pi f C_l}$$

$$C_l = \frac{1}{2\pi f \ |X_C|} = \frac{1}{2\pi \ (60 \ \text{s}^{-1}) \ (0.1090 \times 10^6 \ \Omega\text{-mi})}$$

$$= 2.43 \times 10^{-8} \ \text{F/mi}$$

Substituting the calculated values gives

$$v_w = \frac{1}{\sqrt{L_l C_l}}$$

$$= \frac{1}{\sqrt{\left(1.26 \times 10^{-3} \frac{H}{mi}\right)\left(2.43 \times 10^{-8} \frac{F}{mi}\right)}}$$

$$= \left(1.81 \times 10^5 \frac{mi}{s}\right)\left(1.609 \frac{km}{mi}\right)\left(1000 \frac{m}{km}\right)$$

$$= 2.91 \times 10^8 \text{ m/s}$$

The velocity factor is

$$v_{factor} = \frac{v_w}{C}$$

$$= \frac{2.91 \times 10^8 \frac{m}{s}}{3.00 \times 10^8 \frac{m}{s}}$$

$$= \boxed{0.97}$$

The answer is (D).

4. This is a medium-length transmission line. Regardless of the model chosen,

$$A = 1 + \tfrac{1}{2}YZ$$

$$= 1 + \tfrac{1}{2}\left(7 \times 10^{-6} \frac{\Omega^{-1}}{mi} \angle 90°\right)\left(0.85 \frac{\Omega}{mi} \angle 70°\right)$$

$$= 1 + \left(2.98 \times 10^{-6} \text{ mi}^{-2} \angle 160°\right)(220 \text{ km})^2$$

$$\times \left(\frac{1 \text{ mi}}{1.609 \text{ km}}\right)^2$$

$$= 1\angle 0° + 5.56 \times 10^{-2} \angle 160°$$

$$= (1 + 0j) + (-0.0523 + j0.0190)$$

$$= 0.9477 + j0.0190$$

$$= \boxed{0.948 \angle 1.2°}$$

The type of cable determines the electrical parameters, but these were given. Also, the first term of the parameter, that is, one, can be represented as $1\angle 0°$.

The answer is (B).

5. The sending-end voltage is given by (see Fig. 37.3)

$$V_S = AV_R + BI_R$$

V_R and I_R must be determined. Because the two-part model represents a single phase, the receiving-end phase voltage for a wye-connected load is

$$V_R = \frac{215 \text{ kV}}{\sqrt{3}} = 124 \text{ kV}$$

Because the power factor is given for the load, let V_R be the reference voltage. Thus,

$$V_R = 124 \text{ kV} \angle 0°$$

The receiving-end current is found from

$$P = 3I_R V_R(\text{pf})$$

$$I_R = \frac{P}{3V_R(\text{pf})} = \frac{100 \times 10^6 \text{ W}}{(3)(124 \times 10^3 \text{ V})(0.8)}$$

$$= 336 \text{ A}$$

Because the power factor is 0.8 lagging, the current angle is $-36.8°$.

$$I_R = 336 \text{ A} \angle -36.8°$$

Substituting gives

$$V_S = AV_R + BI_R$$

$$= (0.9\angle 2.0°)(124 \times 10^3 \text{ V} \angle 0°)$$

$$\quad + (150\angle 80°)(336 \text{ A} \angle -36.8°)$$

$$= 1.116 \times 10^5 \text{ V} \angle 2.0° + 5.04 \times 10^4 \text{ V} \angle 43.2°$$

$$= \boxed{153 \text{ kV} \angle 14.5°}$$

Note that the power could have been calculated using the quantities, in which case the current would be a function of the power and the line-to-line voltage, V_{ll}.

$$I_R = \frac{P}{\sqrt{3}V_{ll}(\text{pf})}$$

38 Batteries and Fuel Cells

PRACTICE PROBLEMS

1. A cell that undergoes a chemical reaction in such a manner that it cannot be recharged is called a

    (A) dry cell

    (B) primary cell

    (C) reserve cell

    (D) secondary cell

2. Within a battery, conventional current flows from

    (A) anion to anode

    (B) anion to cation

    (C) anode to cathode

    (D) positive electrode to negative electrode

3. A series of fuel cells is connected to produce a no-load voltage of 320 V. At a rated voltage of 300 V the cells provide 2.0 MW to a DC-AC converter. Assuming the cell can be modeled as a standard battery, what is the total internal resistance of the fuel cells?

    (A) 0.0001 Ω

    (B) 0.003 Ω

    (C) 0.05 Ω

    (D) 3 Ω

4. A wristwatch mercury cell has a capacity of 200 mA·h. How many months can the watch function without replacement if it draws 15 μA?

    (A) 3 mo

    (B) 6 mo

    (C) 12 mo

    (D) 18 mo

5. Increasing the current withdrawn from a battery has what effect on the battery capacity and why?

    (A) increased capacity caused by less internal heat loss

    (B) increased capacity caused by lower internal resistance

    (C) decreased capacity caused by higher load voltage drop

    (D) decreased capacity caused by increased polarization

SOLUTIONS

1. A cell in which the electrodes are consumed and in general cannot be recharged is called a primary cell.

The answer is (B).

2. Within a battery, the conventional current flows from anode to cathode. (The anode is labeled as the positive terminal. Externally, the anode is the terminal into which positive current flows.)

The answer is (C).

3. The circuit is

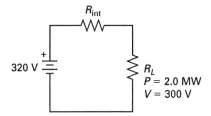

The voltage across the internal resistance is given (20 V). To find the resistance, the current must be determined. From the rated conditions, the load resistance is

$$P = \frac{V^2}{R_L}$$

$$R_L = \frac{V^2}{P} = \frac{(300\ \text{V})^2}{2.0 \times 10^6\ \text{W}} = 0.045\ \Omega$$

The rated current is

$$P = I^2 R_L$$

$$I = \sqrt{\frac{P}{R_L}} = \sqrt{\frac{2.0 \times 10^6\ \text{W}}{0.045\ \Omega}} = 6.67 \times 10^3\ \text{A}$$

Using Ohm's law on the internal resistor gives

$$V = I R_{\text{int}}$$

$$R_{\text{int}} = \frac{V}{I} = \frac{320\ \text{V} - 300\ \text{V}}{6.67 \times 10^3\ \text{A}}$$

$$= \boxed{0.003\ \Omega}$$

The answer is (B).

4. Amp-hour capacity is

$$Ah = It$$

Thus,

$$t = \frac{Ah}{I} = \frac{200 \times 10^{-3} \text{ A·h}}{15 \times 10^{-6} \text{ A}}$$

$$= 1.33 \times 10^4 \text{ h}$$

Assuming 30-day months, the capacity is

$$\text{total months} = \left(1.33 \times 10^4 \text{ h}\right) \left(\frac{1 \text{ d}}{24 \text{ h}}\right) \left(\frac{1 \text{ mo}}{30 \text{ d}}\right)$$

$$= \boxed{18.5 \text{ mo}}$$

The answer is (D).

5. Battery capacity drops as the discharge current increases due to increased losses on the internal resistance and polarization of electrodes, that is, hydrogen bubble insulation of the electrodes.

The answer is (D).

39 Rotating DC Machinery

PRACTICE PROBLEMS

1. A four-pole generator is turned at 3600 rpm. Each of its poles has a flux of 3.0×10^{-4} Wb. The armature has a total of 200 conductors and is simplex lap wound. What average voltage is generated at no load?

2. A DC generator has an armature resistance of 0.31 Ω. The shunt field resistance is 134 Ω. No-load voltage is 121 V at 1775 rpm. Full-load current is 30 A at 110 V. Assume a constant flux. What are the (a) no-load shunt field current, (b) full-load speed, and (c) copper power losses at full load?

3. (a) What is the back emf of a 10 hp shunt motor operating on 110 V terminals if 90 A flow through a 0.05 Ω armature resistance? (b) What is the line current if the field resistance is 60 Ω?

4. The nameplate of a 240 V DC motor states that the full-load line current is 67 A. During a no-load test at rated voltage, 3.35 A armature current and 3.16 A field current are drawn. The no-load armature resistance is 0.207 Ω. The no-load brush drop is 2 V. What are the (a) horsepower and (b) efficiency at full load?

5. A DC generator has a 250 V output at a 50 A load current and a 260 V output at no load. Estimate the armature resistance.

6. A separately excited DC generator is rated at 10 kW, 240 V, and 2800 rpm for an input power of 15 hp. The no-load voltage is 260 V. Determine the (a) armature resistance and (b) mechanical power loss at 2800 rpm.

7. A DC shunt motor is running at rated values when the line voltage suddenly drops by 10%. Determine the resulting changes in input power and motor speed if (a) the load torque is constant and (b) the load torque is proportional to the speed.

8. A DC series motor is running at rated speed, voltage, and current when the load torque increases by 25%. Determine the effects of this change, assuming that the line voltage remains the same.

9. A DC shunt motor takes 200 V and 25 A at 1000 rpm and 5 hp. It has an armature resistance of 0.5 Ω and a shunt field resistance of 100 Ω. Determine the (a) rotational losses and (b) no-load speed at 200 V.

10. A DC shunt motor is rated at 7.5 hp, 60 A, 1000 rpm, and 120 V. Assume the load torque is constant, regardless of speed. What field resistance values are required to change the speed to (a) 1150 rpm and (b) 750 rpm?

SOLUTIONS

1. For z conductors, a parallel armature paths, p poles, a magneetic flux, Φ, and a rotational speed, n,

$$V = \left(\frac{z}{a}\right) p\Phi \left(\frac{n}{60}\right)$$

For simplex lap winding, $a = p$.

$$V = \left(\frac{200}{4}\right)(4)(3.0 \times 10^{-4}\text{ Wb})\left(\frac{3600\ \frac{\text{rev}}{\text{min}}}{60\ \frac{\text{sec}}{\text{min}}}\right)$$

$$= \boxed{3.6\text{ V}}$$

2. (a) The parameters pertain to full load, fl; no load, nl; field, f; and armature, a.

$$I_{f,\text{nl}} = \frac{V_{\text{nl}}}{R_f} = \frac{121\text{ V}}{134\ \Omega}$$

$$= \boxed{0.903\text{ A}}$$

(b) Voltage is proportional to rotational speed.

$$n_{\text{fl}} = \left(\frac{V_{\text{fl}} + R_a\left(I_{\text{fl}} + \frac{V_{\text{fl}}}{R_f}\right)}{V_{\text{nl}} + R_a\left(\frac{V_{\text{nl}}}{R_f}\right)}\right) n_{\text{nl}}$$

$$= \left(\frac{110\text{ V} + (0.31\ \Omega)\left(30\text{ A} + \frac{110\text{ V}}{134\ \Omega}\right)}{121\text{ V} + (0.31\ \Omega)\left(\frac{121\text{ V}}{134\ \Omega}\right)}\right)\left(1775\ \frac{\text{rev}}{\text{min}}\right)$$

$$= \boxed{1750\text{ rev/min} \quad (1750\text{ rpm})}$$

(c) $\quad P = \dfrac{V_{\text{fl}}^2}{R_f} + \left(I_{\text{fl}} + \dfrac{V_{\text{fl}}}{R_f}\right)^2 R_a$

$$= \frac{(110\text{ V})^2}{134\ \Omega} + \left(30\text{ A} + \frac{110\text{ V}}{134\ \Omega}\right)^2(0.31\ \Omega)$$

$$= \boxed{385\text{ W}}$$

3. (a) $\quad E = V_{\text{fl}} - I_{a,\text{load}}R_a$

$$= 110\text{ V} - (90\text{ A})(0.05\ \Omega)$$

$$= \boxed{105.5\text{ V}}$$

(b) $\quad I_{\text{fl}} = \dfrac{V_{\text{fl}}}{R_f} + I_{a,\text{load}} = \dfrac{110\text{ V}}{60\ \Omega} + 90\text{ A}$

$$= \boxed{91.8\text{ A}}$$

4.

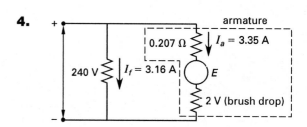

The total no-load power input to the armature is

$$P_a = I_a V = (3.35\text{ A})(240\text{ V})$$

$$= 804\text{ W}$$

The I^2R power loss in the armature resistance is

$$P_{\text{resistance}} = I_a^2 R = (3.35\text{ A})^2(0.207\ \Omega)$$

$$= 2.323\text{ W}$$

The armature brush power loss is

$$P_{\text{brush}} = I_a V_{\text{brush}}$$

$$= (3.35\text{ A})(2\text{ V})$$

$$= 6.7\text{ W}$$

The no-load rotational loss is

$$P_{\text{rotational}} = P_a - P_{\text{resistance}} - P_{\text{brush}}$$

$$= 804\text{ W} - 2.323\text{ W} - 6.7\text{ W}$$

$$= 794.9\text{ W} \quad (795.0\text{ W})$$

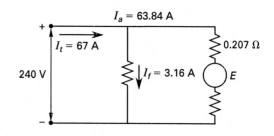

At full load, the total input power is

$$P_{in} = I_{fl} V_{nameplate}$$
$$= (67 \text{ A})(240 \text{ V})$$
$$= 16{,}080 \text{ W}$$

The field current is the same as for the no-load condition.

$$I_f = 3.16 \text{ A}$$

The field power loss is

$$P_f = I_f V = (3.16 \text{ A})(240 \text{ V})$$
$$= 758.4 \text{ W}$$

$$I_a = I_{fl} - I_f = 67 \text{ A} - 3.16 \text{ A}$$
$$= 63.84 \text{ A}$$

$$P_{resistance} = I_a^2 R = (63.84 \text{ A})^2 (0.207 \text{ }\Omega)$$
$$= 843.6 \text{ W}$$

Using the no-load data, the equivalent brush resistance is

$$R_{brush} = \frac{V_{brush}}{I} = \frac{2 \text{ V}}{3.35 \text{ A}} = 0.597 \text{ }\Omega$$

The armature brush power loss is

$$P_{brush} = I_a^2 R_{brush} = (63.84 \text{ A})^2 (0.597 \text{ }\Omega)$$
$$= 2433.1 \text{ W}$$

The rotational loss is the same as for the no-load condition.

$$P_{rotational} = 795.0 \text{ W}$$

The net power to the load is

$$P_{net} = P_{in} - P_f - P_{resistance} - P_{brush} - P_{rotational}$$
$$= 16{,}080 \text{ W} - 758.4 \text{ W} - 843.6 \text{ W}$$
$$\quad - 2433.1 \text{ W} - 795.0 \text{ W}$$
$$= 11{,}249.9 \text{ W}$$

(a) The motor output horsepower is

$$P = \frac{(11{,}249.9 \text{ W})\left(1.341 \dfrac{\text{hp}}{\text{kW}}\right)}{1000 \dfrac{\text{W}}{\text{kW}}} = \boxed{15.09 \text{ hp}}$$

(b) The efficiency is

$$\eta = \frac{P_{net}}{P_{in}} = \frac{11{,}249.9 \text{ W}}{16{,}080 \text{ W}} = \boxed{0.70 \quad (70.0\%)}$$

5.

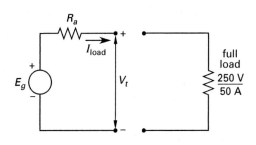

$$V_t = E_g - R_a I_{load}$$

At no load,

$$V_t = E_g = 260 \text{ V}$$

At full load,

$$V_t = E_g - (50 \text{ A}) R_a$$
$$250 \text{ V} = E_g - (50 \text{ A}) R_a$$
$$R_a = \frac{V_{nl} - V_{fl}}{I_{fl}}$$
$$= \frac{260 \text{ V} - 250 \text{ V}}{50 \text{ A}}$$
$$= \boxed{0.2 \text{ }\Omega}$$

6. (a)

$$I_{fl} = \frac{10 \text{ kW}}{240 \text{ V}} = 41.67 \text{ A}$$

$$R_a = \frac{V_{nl} - V_{fl}}{I_{fl}} \quad \text{[see Prob. 39.5]}$$
$$= \frac{260 \text{ V} - 240 \text{ V}}{41.67 \text{ A}}$$
$$= \boxed{0.48 \text{ }\Omega}$$

(b)

$$P_{in} = P_{mech.\ loss} + I_a^2 R_a + p_{out}$$
$$= (15 \text{ hp})\left(\frac{746 \text{ W}}{1 \text{ hp}}\right)$$
$$= 11.190 \text{ kW}$$

$$I_a^2 R_a = (41.67 \text{ A})^2 (0.48 \text{ }\Omega)$$
$$= 0.833 \text{ kW}$$

$$P_{out} = 10 \text{ kW}$$

$$P_{mech.\ loss} = 11.190 \text{ kW} - 10.833 \text{ kW}$$
$$= \boxed{357 \text{ W}}$$

7. For field flux, Φ is proportional to line voltage.

$$\Phi = aV$$

$$\Omega_r = \text{rotor speed} \quad [\text{rad/s}]$$

$$E_g = k\Phi\Omega_r$$

$$= kaV\Omega_r$$

$$T = k\Phi I_a = kaVI_a \quad [\text{torque in N·m}]$$

$$V = E_g + R_a I_a$$

$$P_{\text{in}} \approx VI_a$$

$$P_{\text{out}} \approx T\Omega_r$$

The parameters pertain to the field, f; armature, a; generator, g; and rotor, r.

(a) When V is decreased by 10%, to keep T constant, I_a is increased by approximately 10%, and E_g is decreased by 10% with V.

$$P_{\text{in}} = \boxed{VI_a \approx \text{constant}}$$

$$P_{\text{out}} = P_{\text{in}} - I_a^2 R_a - \text{mech. loss}$$

Because I_a has increased by 10%, I_a^2 is increased by 20%, but $I_a^2 R_a$ is on the order of 5% of the load power, so it only causes a 1% drop in P_{out}.

E_g drops 10% with V, but

$$\boxed{\Omega_r \text{ drops only 1\%.}}$$

(b) For $T \propto \Omega_r$, in a first approximation,

$$E_g \approx V$$

$$E_g = kaV\Omega_r$$

$ka\Omega_r \approx 1$, and there is no change in speed.

In a second approximation,

$$I_a = \frac{V - E_g}{R_a}$$

$$T = kaVI_a$$

The air-gap power is

$$P_{\text{air gap}} = E_g I_a = T\Omega_r$$

$$T = m\Omega_r$$

$$T\Omega_r = m\Omega_r^2$$

$$E_g = kaV\Omega_r$$

$$E_g I_a = \frac{kaV^2\Omega_r(1 - ka\Omega_r)}{R_a} = m\Omega_r^2$$

$$\frac{kaV^2}{R_a} = \frac{m\Omega_r}{1 - ka\Omega_r} = \frac{m}{\frac{1}{\Omega_r} - ka}$$

When V goes down 10%, $1/\Omega_r - ka$ must go up by 21%, so Ω_r must decrease some amount. If under normal conditions $I_a R_a = 0.05V$,

$$I_a R_a = V - E_{g,\text{nl}} = V(1 - ka\Omega_{r,\text{nl}})$$

$$1 - ka\Omega_{r,\text{nl}} = 0.05$$

$$ka = \frac{0.95}{\Omega_{r,\text{nl}}}$$

When V goes down 10%,

$$\frac{m\Omega_r}{1 - ka\Omega_r} \rightarrow (0.81)\left(\frac{m\Omega_{r,\text{nl}}}{1 - ka\Omega_{r,\text{nl}}}\right)$$

$$\frac{\Omega_r}{1 - (0.95)\left(\frac{\Omega_r}{\Omega_{r,\text{nl}}}\right)} = \left(\frac{0.81}{0.05}\right)\Omega_{r,\text{nl}} = 16.2\Omega_{r,\text{nl}}$$

$$\frac{\Omega_r}{\Omega_{r,\text{nl}}} = (16.2)\left(1 - 0.95\left(\frac{\Omega_r}{\Omega_{r,\text{nl}}}\right)\right)$$

$$= \frac{16.2}{1 + (0.95)(16.2)}$$

$$= 0.99$$

$$\boxed{\Omega_r \text{ drops by about 1\%.}}$$

$$I_a R_a = V - E_g = V - ka\Omega_r$$

$$T = m\Omega_r = kaVI_a$$

Eliminating Ω_r gives

$$I_a R_a = \frac{V}{1 + \frac{(kaV)^2}{mR_a}}$$

The original value of $I_a R_a$ is

$$I_{a,\text{nl}} R_a \approx 0.05V_{\text{nl}}$$

V_0 is the initial voltage. Then,

$$1 + \frac{(kaV_{\rm nl})^2}{mR_a} \approx 20$$

When the voltage drops 10%,

$$1 + \frac{\left(ka(0.9)V_{\rm nl}\right)^2}{mR_a} = 16.39$$

The resulting armature IR drop is

$$I_a R_a = \frac{0.9V_{\rm nl}}{16.39} = 0.0549V_{\rm nl}$$

Dropping the common $V_{\rm nl}$ term, the current changes by

$$\frac{I_a R_a - I_{a,{\rm nl}}R_a}{I_{a,{\rm nl}}R_a} = \frac{0.0549 - 0.05}{0.05}$$

$$= 0.098 \text{ pu}$$

The input power changes also.

$$P_{\rm in} = (0.9V_{\rm nl})(1.098I_{a,{\rm nl}}) = 0.988P_{\rm nl}$$

> The power drops by about 1%.

8.

 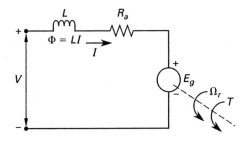

$$V = RI + E_g$$

$$E_g = k\Phi\Omega_r = kLI\Omega_r$$

$$T = k\Phi I = kLI^2$$

If T is up 25%, so is I^2, and $\sqrt{1.25} = 1.11$, so

> I is up 11%.

Using subscript zero for the initial values,

$$E_{g0} = V - I_0 R_a$$

When I increases by 11%, using subscript 1 for this condition,

> E_g decreases by about 1%.

$$E_{g1} = 0.99E_{g0}$$

$$E_{g0} = kLI_0\Omega_{r0}$$

$$E_{g1} = 0.99kLI_0\Omega_{r0}$$

$$= kLI_1\Omega_{r1}$$

$$\Omega_{r1} = \left(\frac{0.99I_0}{I_1}\right)\Omega_{r0}$$

$$= \left(\frac{0.99}{1.11}\right)\Omega_{r0}$$

$$= 0.89\Omega_{r0}$$

> The speed also drops by about 11%.

The output power is

$$P_2 = (1.25T_0)(0.89\Omega_{r0}) = 1.11P_0$$

> The air-gap power increases by 11%.

9. (a)

$$I_t = 25 \text{ A} = I_a + I_f$$

$$I_f = \frac{200 \text{ V}}{100 \text{ }\Omega} = 2 \text{ A}$$

$$I_a = I_t - I_f = 23 \text{ A}$$

$$E_g = 200 \text{ V} - (23 \text{ A})R_a = 188.5 \text{ V}$$

$$P_{\rm air\ gap} = E_g I_a = (188.5 \text{ V})(23 \text{ A})$$

$$= 4335.5 \text{ W}$$

$$P_{\rm load} = (5 \text{ hp})\left(\frac{746 \text{ W}}{1 \text{ hp}}\right) = 3730 \text{ W}$$

$$P_{\rm rotational\ losses} = P_{\rm air\ gap} - P_{\rm load} = 4335.5 \text{ W} - 3730 \text{ W}$$

$$= \boxed{605.5 \text{ W}}$$

(b) The speed, n, is proportional to the counter voltage as seen from

$$E_g = k\Phi n$$

The term $k\Phi$ is determined from the information in Part (a) to be

$$k\Phi = \frac{188.5 \text{ V}}{1000 \ \frac{\rm rev}{\rm min}}$$

Assuming the rotational losses are fixed at 605.5 W,

$$E_g I_a = 605.5 \text{ W}$$

$$I_a = \frac{605.5 \text{ W}}{E_g}$$

The armature current at the specified 200 V is

$$I_a = \frac{200 \text{ V} - E_g}{0.5 \text{ }\Omega}$$

Substitute for the armature current, giving

$$\frac{605.5 \text{ W}}{E_g} = \frac{200 \text{ V} - E_g}{0.5 \text{ }\Omega}$$

Rearranging gives the quadratic equation

$$E_g^2 - 200E_g + 302.75 = 0$$

Solving the quadratic and using the positive sign gives

$$E_g = \frac{-(-200 \text{ V}) \pm \sqrt{(-200 \text{ V})^2 - (4)(1)(302.75 \text{ V}^2)}}{(2)(1)}$$

$$= 198.5 \text{ V}$$

Comparing voltage to rpm ratios gives

$$\frac{n}{198.5 \text{ V}} = \frac{1000 \frac{\text{rev}}{\text{min}}}{188.5 \text{V}}$$

$$n = \boxed{1053 \text{ rev/min} \quad (1053 \text{ rpm})}$$

10.

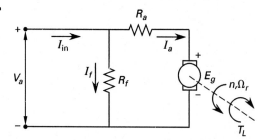

$$P_{\text{in}} = V_A I_{\text{in}} = (60 \text{ A})(120 \text{ V}) = 7200 \text{ W}$$

$$P_{\text{out}} = (7.5 \text{ hp})\left(746 \frac{\text{W}}{\text{hp}}\right) = 5595 \text{ W}$$

Because load torque is constant,

$$P_{\text{out}} = k_T n$$

$$k_T = \frac{5595 \text{ W}}{1000 \frac{\text{rev}}{\text{min}}}$$

$$= 5.595n \quad [n \text{ in rev/min}]$$

The losses are the difference between P_{in} and P_{out}.

$$P_{\text{loss}} = P_{\text{in}} - P_{\text{out}} = 7200 \text{ W} - 5595 \text{ W}$$

$$= 1605 \text{ W}$$

Because no information is given, assign these losses to the rotational losses, the armature losses $I_a^2 R_a$, and the field losses $I_f^2 R_f$, making all three equal in magnitude.

$$I_a^2 R_a = I_f^2 R_f = P_{\text{rotational losses}} = 535 \text{ W}$$

$$I_f^2 R_f = \frac{V_a^2}{R_f} = 535 \text{ W}$$

$$R_f = \frac{(120 \text{ V})^2}{535 \text{ W}} = 26.9 \text{ }\Omega$$

$$I_f = \frac{120 \text{ V}}{26.9 \text{ }\Omega} = 4.46 \text{ A}$$

Therefore,

$$I_a = I_{\text{in}} - I_f = 60 \text{ A} - 4.46 \text{ A} = 55.54 \text{ A}$$

$$R_a = \frac{P_{\text{rotational losses}}}{I_a^2} = \frac{535 \text{ W}}{(55.54 \text{ A})^2} = 0.173 \text{ }\Omega$$

Because of rounding, the powers are recalculated and the remainder assigned to rotational losses.

$$P_{\text{rotational losses}} = P_{\text{loss}} - \frac{V_A^2}{R_f} - R_a I_a^2$$

$$= 1605 \text{ W} - \frac{(120 \text{ V})^2}{26.9 \text{ }\Omega}$$

$$- (0.173 \text{ }\Omega)(55.54 \text{ A})^2$$

$$= 536 \text{ W}$$

The air-gap power must then be

$$P_{\text{air gap}} = P_{\text{out}} + P_{\text{rotational losses}} = 5.595n + 536 \text{ W}$$

The air-gap power is also $E_g I_a$, and E_g is proportional to I_f and n.

$$E_g = k' I_f n = k''\left(\frac{n}{R_f}\right)$$

At $n = 1000$ rpm and $R_f = 26.9 \ \Omega$,

$$E_g = 120 \text{ V} - (55.54 \text{ A})(0.173 \ \Omega)$$

$$= 110.4 \text{ V}$$

$$k'' = \frac{(110.4 \text{ V})(26.9 \ \Omega)}{1000 \ \dfrac{\text{rev}}{\text{min}}}$$

$$= 2.97 \ \frac{\text{V} \cdot \Omega}{\dfrac{\text{rev}}{\text{min}}}$$

$$E_g = 2.97 \left(\frac{n}{R_f}\right)$$

$$I_a = \frac{120 \text{ V} - E_g}{R_a}$$

$$= \frac{120 \text{ V} - (2.97)\left(\dfrac{n}{R_f}\right)}{0.173 \ \Omega}$$

$$= 694 \text{ A} - (17.2)\left(\frac{n}{R_f}\right)$$

$$P_{\text{air gap}} = E_g I_a$$

$$= (2061)\left(\frac{n}{R_f}\right) - (51.08)\left(\frac{n}{R_f}\right)^2$$

Equating with the mechanical side,

$$(51.08)\left(\frac{n}{R_f}\right)^2 + \left(5.595 - \frac{2061}{R_f}\right)n + 536 \text{ W} = 0$$

$$0 = R_f^2 - \left(\frac{2061n}{5.595n + 536 \text{ W}}\right)(R_f) + \frac{51.08n^2}{5.595n + 536 \text{ W}}$$

(a) For $n = 1150$ rpm,

$$R_f^2 - (340.0 \ \Omega)R_f + 9692 \ \Omega^2 = 0 \ \Omega^2$$

$$\boxed{R_f = 31.4 \ \Omega}$$

(b) For $n = 750$ rpm,

$$R_f^2 - (326.6 \ \Omega)R_f + 6072 \ \Omega^2 = 0 \ \Omega^2$$

$$\boxed{R_f = 19.8 \ \Omega}$$

40 Rotating AC Machinery

PRACTICE PROBLEMS

1. An old single-phase generator has four poles. The armature is simplex lap wound and has a total of 240 armature conductors. The effective armature length and radius are 16.7 cm and 5 cm, respectively. The armature rotates at 1800 rpm. (a) If the uniform magnetic flux density per pole is 1.0 T, what is the maximum voltage induced? (b) What is the horsepower of the driving motor if the rated output is 1 kW and the conversion efficiency is 90%?

2. (a) What is the rotational speed of a 24-pole alternator that produces a 60 Hz sinusoid? (b) If the effective emf is 2200 V and each phase coil has 20 turns in series, what is the maximum flux?

3. Calculate the full-load phase current drawn by a 440 V (rms), 20 hp (per phase) induction motor having a full-load efficiency of 86% and a full-load power factor of 76%.

4. A 200 hp, three-phase, four-pole, 60 Hz, 440 V (rms) squirrel-cage induction motor operates at full load with an efficiency of 85%, a power factor of 91%, and 3% slip. Find the (a) speed in rpm, (b) torque developed, and (c) line current.

5. A factory's induction motor load draws 550 kW at 82% power factor. What size synchronous motor is required to carry 250 hp and raise the power factor to 95%? The line voltage is 220 V (rms).

6. The nameplate of an induction motor lists 960 rpm as the full-load speed. For what frequency was the motor designed?

7. A three-phase, 440 V (line-to-line) synchronous motor has a synchronous reactance (X_s) of 20 Ω and an armature resistance of 0.5 Ω per phase. The synchronous speed is 1800 rpm. The no-load field current is adjusted for minimum armature current at a field current (I_f) of 5 A DC, at which time the input power is 900 W. This machine is to provide 10 kVAR of power factor correction when running lightly loaded. Determine the rotor current required.

8. A 440 V, three-phase synchronous motor is driving a 45 hp pump. Assume the motor is 93% efficient. The rotor current is adjusted until the line current is at a minimum, occurring when the rotor current is 8.2 A DC. Determine the rotor current necessary to provide a leading power factor of 0.8.

9. A 50 hp, 50 kVA, 1746 rpm squirrel-cage induction motor has phase windings rated at 440 V. With the windings connected in a delta configuration, a blocked rotor test is carried out under rated current conditions. The result in per-phase volt-amperes is found to be $P + jQ = 1097 + j2791$ VA. This machine is fed from a three-phase, 50 kVA, 440 V transformer that has 10% reactance per phase (on its own base). Determine the dip in the line voltage when the motor is started. The transformer can be modeled as a wye connection of 254 V sources in series with the 10% reactance on each phase.

10. A 50 hp, 50 kVA, 1746 rpm squirrel-cage induction motor has phase windings rated at 440 V and is normally connected in delta. Its starting current is 367 A and 38.3 kW per phase when started directly across the line, and under those conditions it develops a starting torque of 190 ft-lbf. A blocked rotor test at the rated current results in a voltage of 78.6 V and a power of 1.1 kW per phase. A no-load test at the rated voltage results in a current of 25.9 A and 1.1 kW per phase. If this machine is started in the wye configuration, determine its starting current.

SOLUTIONS

1. (a) The maximum voltage is

$$V_m = \left(\frac{\pi}{2}\right) V_{\text{ave}}$$

In terms of the rotational speed, n, the maximum voltage is given in Table 40.1, using the total number of poles, p; the number, N, of series armature paths; area, A; and magnetic flux density, B.

$$V_m = \frac{\pi p n N A B}{60}$$

$$= \pi p \left(\frac{n}{60}\right)\left(\frac{z}{a}\right) A B$$

For a simplex lap-wound armature, $a = p$.

$$V_m = (\pi)(4)\left(\frac{1800 \; \frac{\text{rev}}{\text{min}}}{60 \; \frac{\text{s}}{\text{min}}}\right)\left(\frac{240}{4}\right)$$

$$\times \left((0.10 \text{ m})(0.167 \text{ m})\right)\left(1.0 \; \frac{\text{Wb}}{\text{m}^2}\right)$$

$$= \boxed{377.8 \text{ V}}$$

(b) $$P_{\text{in}} = \frac{P_{\text{out}}}{\text{efficiency}} = \frac{(1 \text{ kW})\left(\dfrac{1 \text{ hp}}{0.7457 \text{ kW}}\right)}{0.9}$$

$$= \boxed{1.49 \text{ hp}}$$

2. (a) $n = 120\dfrac{f}{p}$

$$n = \left(2 \; \frac{\text{pole} \cdot \text{rev}}{\text{cycle}}\right)\left(60 \; \frac{\text{s}}{\text{min}}\right)\left(\frac{f}{p}\right)$$

$$= \left(2 \; \frac{\text{pole} \cdot \text{rev}}{\text{cycle}}\right)\left(60 \; \frac{\text{s}}{\text{min}}\right)\left(\frac{60 \text{ Hz}}{24 \text{ poles}}\right)$$

$$= \boxed{300 \text{ rpm}}$$

(b) $$BA = \frac{\sqrt{2} \, V_{\text{eff}}}{N\omega} = \frac{\sqrt{2}(2200 \text{ V})}{(20)(2\pi)(60 \text{ Hz})}$$

$$= \boxed{0.413 \text{ Wb}}$$

3. $$I_p = \frac{P_p}{\eta V_p \cos\phi}$$

$$= \frac{(20 \text{ hp})\left(\dfrac{0.7457 \times 10^3 \text{ W}}{1 \text{ hp}}\right)}{(0.86)(440 \text{ V})(0.76)}$$

$$= \boxed{51.86 \text{ A}}$$

4. (a) $n_r = \left(\dfrac{(2)\left(60 \; \dfrac{\text{s}}{\text{min}}\right)f}{p}\right)(1-s)$

$$= \left(\frac{(2)\left(60 \; \frac{\text{s}}{\text{min}}\right)(60 \text{ Hz})}{4}\right)(1-0.03)$$

$$= \boxed{1746 \text{ rpm}}$$

(b) $$T = \frac{P}{\Omega} = \frac{(200 \text{ hp})\left(550 \; \dfrac{\frac{\text{ft-lbf}}{\text{s}}}{\text{hp}}\right)}{2\pi\left(1746 \; \dfrac{\text{rev}}{\text{min}}\right)\left(\dfrac{1 \text{ min}}{60 \text{ s}}\right)}$$

$$= \boxed{602 \text{ ft-lbf}}$$

(c) $$I_l = \left(\frac{1}{\sqrt{3}}\right)\left(\frac{P}{\eta V_l \cos\phi}\right)$$

$$= \left(\frac{1}{\sqrt{3}}\right)\left(\frac{(200 \text{ hp})\left(0.7457 \times 10^3 \; \frac{\text{W}}{\text{hp}}\right)}{(0.85)(440 \text{ V})(0.91)}\right)$$

$$= \boxed{253 \text{ A}}$$

5.

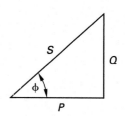

The original power angle is

$$\phi_1 = \arccos 0.82 = 34.92°$$

$$P_1 = 550 \text{ kW}$$

$$Q_i = P_1 \tan\phi_1$$

$$= (550 \text{ kW})(\tan 34.92°)$$

$$= 384.0 \text{ kVAR}$$

The new conditions are

$$P_2 = (250 \text{ hp}) \left(0.7457 \frac{\text{kW}}{\text{hp}} \right)$$

$$= 186.4 \text{ kW}$$

$$\phi_f = \arccos 0.95 = 18.19°$$

Because both motors perform real work,

$$P_f = P_1 + P_2 = 550 \text{ kW} + 186.4 \text{ kW}$$

$$= 736.4 \text{ kW}$$

The new reactive power is

$$Q_f = P_f \tan \phi_f = (736.4 \text{ kW})(\tan 18.19°)$$

$$= 242.0 \text{ kVAR}$$

The change in reactive power is

$$\Delta Q = Q_i - Q_f$$
$$= 384.0 \text{ kVAR} - 242.0 \text{ kVAR} = 142 \text{ kVAR}$$

Synchronous motors used for power factor correction are rated by apparent power.

$$S = \sqrt{(\Delta P)^2 + (\Delta Q)^2}$$

$$= \sqrt{(186.4 \text{ kW})^2 + (142 \text{ kVAR})^2}$$

$$= \boxed{234.3 \text{ kVA}}$$

6.
$$f = \frac{pn_s}{(2)\left(60 \frac{\text{s}}{\text{min}}\right)}$$

$$= \frac{pn}{(2)\left(60 \frac{\text{s}}{\text{min}}\right)(1-s)}$$

The slip and number of poles are unknown. Assume $s = 0$ and $p = 4$.

$$f = \frac{(4)\left(960 \frac{\text{rev}}{\text{min}}\right)}{(2)\left(60 \frac{\text{s}}{\text{min}}\right)} = 32 \text{ Hz}$$

This is not close to anything in commercial use. Try $p = 6$.

$$f = \frac{(6)\left(960 \frac{\text{rev}}{\text{min}}\right)}{(2)\left(60 \frac{\text{s}}{\text{min}}\right)} = 48 \text{ Hz}$$

With a 4% slip, $f = 50$ Hz.

$$\boxed{50 \text{ Hz (European)}}$$

7. The required volt-amps per phase are determined by

$$\frac{Q}{3 \text{ phases}} = V_a I_a = \frac{V_a (E_g - V_a)}{X_s}$$

$$E_g = \left(\frac{Q}{3 \text{ phases}} \right) \left(\frac{X_s}{V_a} \right) + V_a$$

$$= \left(\frac{10^4 \text{ VA}}{3 \text{ phases}} \right) \left(\frac{20 \text{ }\Omega}{254 \text{ V}} \right) + \frac{440 \text{ V}}{\sqrt{3}}$$

$$= 516 \text{ V}$$

$$I_f = (516 \text{ V}) \left(\frac{5 \text{ A}}{254 \text{ V}} \right)$$

$$I_r = E_g \left(\frac{I_f}{V_a} \right)$$

$$= \boxed{10.16 \text{ A}}$$

8.
$$P_{\text{in}} = \frac{P_{\text{out}}}{\text{efficiency}}$$

$$= \frac{(45 \text{ hp})\left(746 \frac{\text{W}}{\text{hp}} \right)}{0.93} = 36.1 \text{ kW}$$

At the minimum line current, the current is in phase with the voltage.

$$I = \frac{\dfrac{36.1 \times 10^3 \text{ W}}{3 \text{ phases}}}{\dfrac{440 \text{ V}}{\sqrt{3}}} = 47.37 \text{ A}$$

$$Z_{\text{base}} = \frac{254 \text{ V}}{47.37 \text{ A}} = 5.36 \text{ }\Omega$$

A synchronous reactance of 100% is common and is therefore assumed.

$$X_s = 5.36 \text{ }\Omega$$

$$I_a X_s = (5.36 \text{ }\Omega)(47.37 \text{ A}) = 254 \text{ V}$$

The emf is

$$E_g = V_a - jX_s I_a = (254 \text{ V})(1-j)$$

$$|E_g| = 359 \text{ V}$$

Power— Machinery

The rotor current is given as $I_r = 8.2$ A.

To provide a leading power factor of 0.8, it is necessary to retain the same power.

$$|I_a| \cos\theta = 47.37 \text{ A}$$

$$|I_a| = \frac{47.37 \text{ A}}{0.8} = 59.2 \text{ A}$$

$$I_a = (59.2 \text{ A})(\cos\theta + j \sin\theta)$$

$$= 47.37 \text{ A} + j35.52 \text{ A}$$

$$= 59.2 \text{ A}\angle 36.87°$$

$$E_g = 254 \text{ V} - (j5.36 \ \Omega)I_a$$

$$= 512 \text{ V}\angle -29.74°$$

Using a ratio gives

$$I_r = \left(\frac{8.2 \text{ A}}{359 \text{ V}}\right)(512 \text{ V})$$

$$= \boxed{11.7 \text{ A}}$$

9. The bases for the motor and transformer are the same.

$$I_{\text{rated}} = \frac{\dfrac{50 \text{ kVA}}{3 \text{ phases}}}{254 \text{ V}} = 65.62 \text{ A}$$

$$Z_{\text{base}} = \frac{254 \text{ V}}{65.62 \text{ A}} = 3.87 \ \Omega$$

From the blocked rotor test,

$$P + jQ = I^2(R_e + jX_e)$$

$$1097 \text{ W} + j2791 \text{ VAR} = (65.62 \text{ A})^2(R_e + jX_e)$$

$$R_e + jX_e = 0.2548 + j0.6482 \ \Omega$$

On a per-unit basis,

$$R_e + jX_e = 0.0658 + j0.1675 \text{ pu}$$

The per-unit circuit for starting is

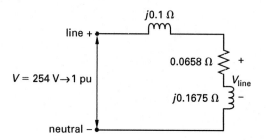

By voltage division,

$$|V_{\text{line}}| = \left|\frac{0.0658 + j0.1675 \ \Omega}{0.0658 + j0.2675 \ \Omega}\right| = 0.653 \text{ pu}$$

$$\text{dip} = \boxed{1 - 0.653 \text{ pu } (35\%)}$$

10. The impedance for wye starting is three times that for delta starting, so the current will be only one-third the delta current.

$$I_{\text{starting}} = \frac{367 \text{ A}}{3} = \boxed{122.3 \text{ A}}$$

41 Lightning Protection and Grounding

PRACTICE PROBLEMS

1. What range of voltage does lightning span?
(A) 1 MV to 2 MV
(B) 2 MV to 5 MV
(C) 5 MV to 10 MV
(D) 10 MV to 1000 MV

2. What range of current does lightning span?
(A) 100 A to 1000 A
(B) 1000 A to 10,000 A
(C) 1000 A to 100,000 A
(D) 1000 A to 200,000 A

3. What is the name for an insulation level specified in terms of the crest value of the standard lightning impulse?
(A) BIL
(B) lightning impulse level
(C) TIL
(D) withstand voltage

4. A power system designed for what voltage is considered protected from direct lightning strikes by its own insulation?
(A) 15 kV
(B) 230 kV
(C) 330 kV
(D) none of the above

5. An arrester provides a protection level of 38 kV. If the minimum protective margin allowed by standards is 20%, is equipment with a 95 kV withstand-voltage adequately protected?

SOLUTIONS

1. Lightning voltage varies widely but is in the range of 10 MV to 1000 MV.

The answer is (D).

2. The current ranges from 1000 A to 200,000 A.

The answer is (D).

3. The voltage level that insulation is able to withstand, specified in terms of a standard lightning impulse, is called the basic lightning impulse level (BIL).

The answer is (A).

4. Systems of 230 kV or less cannot be insulated against the overvoltage caused by direct lightning strikes and must use a combination of shielding and grounding.

The answer is (C).

5. The protective margin is

$$\text{PM} = \frac{V_w - V_p}{V_p}$$

$$= \frac{95 \text{ kV} - 38 \text{ kV}}{38 \text{ kV}}$$

$$= 1.5 \ (150\%) > 20\%$$

The equipment is adequately protected.

42 Illumination

PRACTICE PROBLEMS

1. What is the wavelength of a 60 Hz signal (using vacuum as the reference)?
- (A) 0.016 m
- (B) 60 m
- (C) 5.0×10^6 m
- (D) 3.0×10^8 m

2. What is the energy contained in a single quantum of visible light at a wavelength of 5.55×10^{-7} m?
- (A) 3.7×10^{-40} J
- (B) 3.6×10^{-19} J
- (C) 5.6×10^{-7} J
- (D) 170 J

3. Which type of light may be used to destroy bacteria?
- (A) infrared
- (B) laser
- (C) visible
- (D) ultraviolet

4. Which of the following statements about visible light is NOT true?
- (A) the index of refraction is 1.5 in air
- (B) visible light consists of many colors
- (C) the average frequency is approximately 5.4×10^{14} Hz
- (D) the wavelengths are approximately 380–780 nm

5. Kirchhoff's law of thermal radiation specifies that the emissitivity, ε, equals the spectral absorption, α, in
- (A) a blackbody
- (B) a greybody
- (C) a selective radiator
- (D) none of the above

6. What is the maximum wavelength expected for a light source at 3000K?
- (A) 9.7×10^{-7} m
- (B) 3.3×10^{-4} m
- (C) 2.9×10^{-3} m
- (D) 3.0×10^{-3} m

7. What wavelength of light is emitted if an electron experiences a 2.2 V transition in a semiconductor?
- (A) 220 nm
- (B) 300 nm
- (C) 555 nm
- (D) 560 nm

8. The maximum lumens available per watt at a wavelength of 555 nm is _____ lm, though an ideal white source provides a maximum of _____ lm per watt.
- (A) 555, 220
- (B) 683, 220
- (C) 683, 683
- (D) 683, 432

9. The illuminance required at desktop level for a given task is 500 lx. Manufacturer data for the light source indicates an illuminance of 3500 lx at 1 m. The situation is as shown.

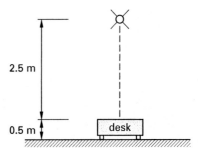

What is the illuminance level at the desk top?
- (A) 80 lx
- (B) 560 lx
- (C) 1400 lx
- (D) 14 000 lx

10. Consider the lighting configuration shown. The light sources all have the same intensity, $I = 500$ cd. What is the contribution of light source 2 at point A?

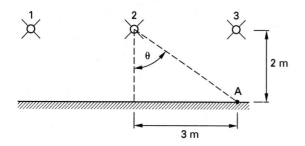

(A) 21 lx

(B) 42 lx

(C) 70 lx

(D) 130 lx

SOLUTIONS

1. The wavelength (with an index of refraction equal to one—that of a vacuum) is derived from the relation of the speed of light, c, to its wavelength, λ, and its frequency, ν.

$$c = \lambda \nu$$

$$\lambda = \frac{c}{\nu} = \frac{3 \times 10^8 \, \frac{\text{m}}{\text{s}}}{60 \, \text{s}^{-1}}$$

$$= \boxed{5 \times 10^6 \, \text{m}}$$

The wavelength is 5000 km or approximately 3100 mi. Because the index of refraction in air is nearly equal to the refraction index in a vacuum, this is also the wavelength in air. Extremely low frequency (ELF) waves (such as a 60 Hz signal) travel in the cavity between the Earth's surface and the ionosphere.

The answer is (C).

2. The human retina is most sensitive to visible light at 555 nm (5.55×10^{-7} m). Using Planck's constant, h, the energy of a single quantum of light at this wavelength is

$$E = h\nu = \frac{hc}{\lambda}$$

$$= \frac{(6.6256 \times 10^{-34} \, \text{J·s}) \left(3 \times 10^8 \, \frac{\text{m}}{\text{s}} \right)}{5.55 \times 10^{-7} \, \text{m}}$$

$$= \boxed{3.58 \times 10^{-19} \, \text{J} \quad (3.6 \times 10^{-19} \, \text{J})}$$

The answer is (B).

3. Ultraviolet light causes numerous biological effects, one of which is the destruction of various forms of bacteria.

The answer is (D).

4. The wavelengths of visible light are approximately 380–780 nm. Visible light is polychromatic, that is, it consists of many colors. The human retina is most sensitive to light at 555 nm. The index of refraction in air is approximately one. (The actual difference from one is so slight that it is unimportant in most calculations.)

The answer is (A).

5. A blackbody radiator has the property that

$$\varepsilon = \alpha = 1$$

The answer is (A).

6. According to the Wein displacement law, the maximum wavelength depends only on the temperature, and is given by

$$\lambda_{\max} = \frac{b}{T}$$

$$= \frac{2.8978 \times 10^{-3} \text{ m·K}}{3000\text{K}}$$

$$= \boxed{9.7 \times 10^{-7} \text{ m}}$$

The constant b is determined by experiment.

The answer is (A).

7. The wavelength of emitted radiation in terms of the potential difference is

$$\lambda_{\text{nm}} = \frac{1239.76 \text{ nm·V}}{\Delta V}$$

$$= \frac{1239.76 \text{ nm·V}}{2.2 \text{ V}}$$

$$= \boxed{563 \text{ nm} \quad (560 \text{ nm})}$$

The answer is (D).

8. The theoretical maximum number of lumens available per watt of energy is 683 lm. Because of inefficiencies and unavoidable losses, an ideal white source can only provide 220 lm.

The answer is (B).

9. The inverse square law gives the illuminance from a point source at a straight-line distance, r, from the source as

$$E_1 r_1^2 = E_2 r_2^2$$

$$E_2 = \frac{E_1 r_1^2}{r_2^2}$$

$$= \frac{(3500 \text{ lx})(1 \text{ m})^2}{(2.5 \text{ m})^2}$$

$$= \boxed{560 \text{ lx}}$$

The answer is (B).

10. By the cosine-cubed law, the illuminance E, for a light source of intensity I, at point A, is

$$E = \frac{I \cos^3 \theta}{h^2} \qquad \text{[I]}$$

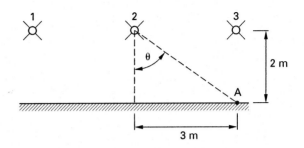

The variable h is the height. The only unknown is the angle θ.

$$\theta = \arctan \frac{\text{opposite}}{\text{adjacent}}$$

$$= \arctan \frac{3 \text{ m}}{2 \text{ m}}$$

$$= 56.3°$$

Substitute the value of θ into Eq. [I].

$$E = \frac{I \cos^3 \theta}{h^2}$$

$$= \frac{(500 \text{ cd})(\cos^3 56.3°)}{(2 \text{ m})^2}$$

$$= \boxed{21.35 \frac{\text{cd}}{\text{m}^2} \quad (21 \text{ lx})}$$

The answer is (A).

Power—
Special Apps

43 Measurement and Instrumentation

PRACTICE PROBLEMS

1. What instrument term can be defined as a measure of the reproducibility of observations?

(A) accuracy

(B) error

(C) precision

(D) resolution

2. A digital voltmeter is designed for a maximum ambient temperature of 65°C. On the 10 kHz scale, what is the expected noise voltage in a 100 kΩ series resistor?

(A) 4.3 nV

(B) 0.14 μV

(C) 1.9 μV

(D) 4.3 μV

3. What is the average value of the voltage signal $v(t) = 5 \sin 377t$?

(A) 0.0 V

(B) 3.2 V

(C) 3.5 V

(D) 5.6 V

4. The voltage signal in Prob. 3 is rectified and measured by a d'Arsonval meter movement. What is the voltage reading?

(A) 0.0 V

(B) 3.2 V

(C) 3.5 V

(D) 5.6 V

5. The manufacturer's data sheet lists the sensitivity of a DC voltmeter as 1000 Ω/V. What current is required for full-scale meter deflection?

(A) 1.0 nA

(B) 1.0 μA

(C) 1.0 mA

(D) 1.0 A

6. A DC voltmeter with a coil resistance of 150 Ω is to be designed with a sensitivity of 10,000 Ω/V and a voltage range of 0 V to 100 V. What type and what magnitude of resistance are required?

(A) parallel; 10 kΩ

(B) parallel; 1 MΩ

(C) series; 10 kΩ

(D) series; 1 MΩ

7. A d'Arsonval movement with a coil resistance of 150 Ω is to be designed for use as an ammeter. What is the full-scale deflection current?

(A) 0.3 mA

(B) 0.3 A

(C) 3.0 A

(D) none of the above

8. A DC voltmeter with 20,000 Ω of internal resistance measures the voltage across a resistance as 10.0 V. The current through the parallel combination of the resistor and voltmeter is measured as 1.00 mA by an ammeter with 100 Ω of internal resistance. The resistor is fed from a source with a Thevenin resistance of 600 Ω. Determine the (a) resistance of the resistor and (b) current that will flow in the resistor if the meters are removed from the circuit.

9. A voltmeter is designed to measure voltages in the full-scale ranges of 3, 10, 30, and 100 V_{DC}. The meter movement has an internal resistance of 50 Ω and a full-scale current of 1 mA. Using a four-contact multiposition switch, design the voltmeter and specify the four resistor values used.

10. A digital meter uses the number system base b and has n digits. The noise is one-half the place value of the least significant digit. (a) Determine the signal-to-noise ratio for the case where the reading is half the full-scale value. (b) Obtain the signal-to-noise ratios for the following cases.

n	b	S/N (dB)
3	10	$20 \log (S/N)$
4	8	
8	2	
3	16	

SOLUTIONS

1. Precision is the range of values of a set of measurements. Therefore, it is a measure of the reproducibility of observations.

The answer is (C).

2. The thermal noise voltage is

$$V_{\text{noise}} = \sqrt{4\kappa T R(\text{BW})}$$

$$= \sqrt{\begin{array}{c} (4)\left(1.3805 \times 10^{-23} \dfrac{\text{J}}{\text{K}}\right)(338 \text{ K}) \\[2mm] \times (100 \times 10^3 \ \Omega)\left(10 \times 10^3 \ \dfrac{1}{\text{s}}\right) \end{array}}$$

$$= \boxed{4.3 \times 10^{-6} \text{ V} \quad (4.3 \ \mu\text{V})}$$

The answer is (D).

3. The average value of a sinusoid is zero.

The answer is (A).

4. A d'Arsonval meter responds to average values. The average value of a rectified sinusoid is

$$V_{\text{ave}} = \left(\frac{2}{\pi}\right) V_p$$

$$= \left(\frac{2}{\pi}\right)(5 \text{ V})$$

$$= \boxed{3.18 \text{ V}}$$

The answer is (B).

5. The sensitivity is defined as

$$\text{sensitivity} = \frac{1}{I_{\text{fs}}}$$

$$I_{\text{fs}} = \frac{1}{\text{sensitivity}} = \frac{1}{1000 \ \dfrac{\Omega}{\text{V}}}$$

$$= \boxed{10^{-3} \text{ A}}$$

The answer is (C).

6. The voltmeter requires an external series resistance given by

$$\frac{1}{I_{\text{fs}}} = \frac{R_{\text{ext}} + R_{\text{coil}}}{V_{\text{fs}}}$$

$$R_{\text{ext}} = V_{\text{fs}}\left(\frac{1}{I_{\text{fs}}}\right) - R_{\text{coil}}$$

$$= (100 \text{ V})\left(10,000 \ \frac{\Omega}{\text{V}}\right) - 150 \ \Omega$$

$$= \boxed{0.999 \text{ M}\Omega \quad (1 \text{ M}\Omega)}$$

The answer is (D).

7. For ammeters, the standard design voltage is 50 mV. The full-scale deflection current is

$$I_{\text{fs}} = \frac{V}{R_{\text{coil}}} = \frac{50 \times 10^{-3} \text{ V}}{150 \ \Omega}$$

$$= \boxed{3.33 \times 10^{-4} \text{ A} \quad (0.3 \text{ mA})}$$

The answer is (A).

8.

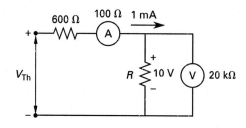

(a) Assuming the meters are perfectly accurate,

$$V_{\text{Th}} = 10 \text{ V} + (0.001 \text{ A})(700 \ \Omega) = 10.7 \text{ V}$$

With 1 mA flowing into 10 V, the equivalent load resistance is 10 V/1 mA = 10 kΩ, so

$$10 \text{ k}\Omega = \frac{(20 \text{ k}\Omega)R}{20 \text{ k}\Omega + R}$$

$$(20 \text{ k}\Omega + R)(10 \text{ k}\Omega) = (20 \text{ k}\Omega)R$$

$$200 \text{ M}\Omega + (10 \text{ k}\Omega)R = (20 \text{ k}\Omega)R$$

$$200 \text{ M}\Omega = (10 \text{ k}\Omega)R$$

$$R = \boxed{20 \text{ k}\Omega}$$

(b) Removing the meters, the current flowing is

$$I = \frac{10.7 \text{ V}}{20{,}600 \ \Omega} = \boxed{0.52 \text{ mA}}$$

9. The meter configuration is shown.

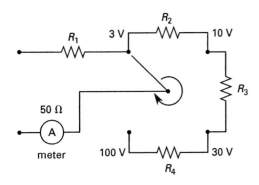

With the 1 mA full-scale current,

$$3 \text{ V:} \quad 50 + R_1 = \frac{3 \text{ V}}{1 \text{ mA}} = 3000 \ \Omega$$

$$R_1 = \boxed{2950 \ \Omega}$$

$$10 \text{ V:} \quad R_2 = \frac{10 \text{ V} - 3 \text{ V}}{1 \text{ mA}}$$

$$= \frac{7 \text{ V}}{1 \text{ mA}} = \boxed{7000 \ \Omega}$$

$$30 \text{ V:} \quad R_3 = \frac{30 \text{ V} - 10 \text{ V}}{1 \text{ mA}}$$

$$= \frac{20 \text{ V}}{1 \text{ mA}} = \boxed{20{,}000 \ \Omega}$$

$$100 \text{ V:} \quad R_4 = \frac{100 \text{ V} - 30 \text{ V}}{1 \text{ mA}}$$

$$= \frac{70 \text{ V}}{1 \text{ mA}} = \boxed{70{,}000 \ \Omega}$$

10. (a) Full-scale is 999_{10} for the first case. The noise is 0.5. At half-scale the signal is 500_{10}, so the signal-to-noise ratio is $500/0.5 = 10^3$ or

$$\text{S/N}|_{\text{dB}} = 20 \log 10^3 = \boxed{60 \text{ dB}}$$

(b) For the octal, full-scale in base 10 is

$$(7)(8)^3 + (7)(8)^2 + (7)(8)^1 + 7 = 4095$$

Half-scale is 2048.

$$\text{S/N} = 20 \log \frac{2048}{0.5} = \boxed{72.2 \text{ dB}}$$

For the binary, full-scale in base 10 is

$$(2)^7 + (2)^6 + (2)^5 + (2)^4 + (2)^3 + (2)^2 + (2)^1 + 1 = 255$$

Half-scale is 128.

$$\text{S/N} = 20 \log \frac{128}{0.5} = \boxed{48.2 \text{ dB}}$$

For the hexadecimal, full-scale is

$$(15)(16)^2 + (15)(16)^1 + 15 = 4095$$

For half-scale, the S/N ratio is the same as the octal half-scale.

$$\text{S/N} = 20 \log \frac{2048}{0.5} = \boxed{72.2 \text{ dB}}$$

44 Electronic Components and Circuits

PRACTICE PROBLEMS

1. Intrinsic germanium has a concentration of electrons of 2.5×10^{13} cm^{-3}. Determine the number of holes.
- (A) 1.2×10^{13} cm^{-3}
- (B) 1.5×10^{13} cm^{-3}
- (C) 2.0×10^{13} cm^{-3}
- (D) 2.5×10^{13} cm^{-3}

2. What type of semiconductor material results if phosphor is the only impurity?
- (A) acceptor
- (B) donor
- (C) n-type
- (D) p-type

3. A silicon semiconductor is doped with 10^{15} atoms/cm^3 of phosphorus. What is the concentration of electrons?
- (A) 1.6×10^{-5} cm^{-3}
- (B) 2.6×10^{5} cm^{-3}
- (C) 0.5×10^{15} cm^{-3}
- (D) 1.0×10^{15} cm^{-3}

4. A silicon semiconductor is doped with a concentration of 10^{15} atoms cm^{-3} of phosphorus. What is the concentration of holes?
- (A) 1.6×10^{-5} cm^{-3}
- (B) 2.6×10^{5} cm^{-3}
- (C) 0.5×10^{15} cm^{-3}
- (D) 1.0×10^{15} cm^{-3}

5. A germanium semiconductor is doped with an impurity having three valence electrons. What are the majority carriers?
- (A) conduction band electrons
- (B) conduction band holes
- (C) valence band electrons
- (D) valence band holes

6. Consider the common base (CB) circuit shown.

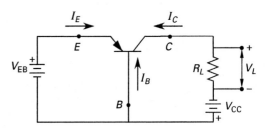

If the collector supply voltage has a magnitude of 1 V and the load is 50 Ω, determine and draw the load line.

7. A discrete silicon semiconductor diode is forward biased at 0.75 V. What is the approximate current?
- (A) 2.0 mA
- (B) 3.0 nA
- (C) 1.5 A
- (D) 3.0 MA

8. The compound GaP, with an energy gap of 2.26 eV, is to be used in an LED. What is the expected wavelength of the emitted photons?
- (A) 8.79×10^{-26} m
- (B) 5.49×10^{-7} m
- (C) 3.42×10^{12} m
- (D) 5.46×10^{14} m

9. At 20°C, a germanium diode passes 70 μA when the reverse bias is -1.5 V. (a) What is the saturation current? (b) What is the current that flows when a forward bias of $+0.2$ V is applied at 20°C, and (c) at 40°C?

10. A voltage of $20 + 5\sqrt{2}\sin 60t$ V is applied to the circuit shown. The diode characteristics are a static forward resistance, r_f, of 120 Ω and a dynamic resistance, r_p, of 100 Ω. What is the voltage across the inductance?

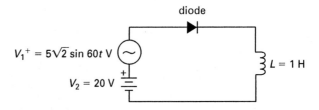

11. At 25°C, a germanium diode shows a saturation current of 100 μA. (a) What current is expected at 100°C, when the diode becomes "useless"? (b) What current is expected at 0°C? (c) At 25°C, what current is predicted for a voltage of -0.5 V?

12. The forward resistance of the diode shown is $R_f = 5$ kΩ. (a) What current is expected in the load resistor $R_L = 5$ kΩ if $V = 100\sin\omega t$ V? (b) Plot the current through and voltage across the external load resistance versus ωt for $0 < \omega t < 2\pi$.

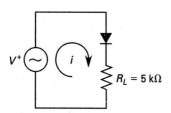

13. A silicon *npn* transistor is used in the circuit shown. The transistor parameters are $I_{\rm CBO} = 10^{-6}$A and $\alpha_0 = 0.99$. The base-emitter voltage is 0.6 V. What is the quiescent collector current, I_C?

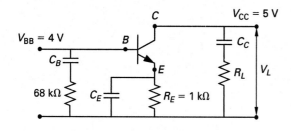

14. A silicon *npn* transistor is shown. Use $\alpha = 0.95$, $I_{\rm CO} = 10^{-6}$ A, and $V_{\rm BE} = 0.6$ V. What current, I_C, will result from this arrangement?

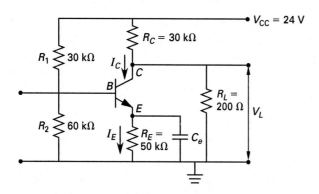

15. The equivalent circuit h-parameters for a bipolar junction transistor operating near the center of its active region in a common emitter configuration are $h_{\rm ie} = 500$ Ω and $h_{\rm fe} = 100$. Determine the values of (a) $h_{\rm ib}$, (b) $h_{\rm fc}$, (c) $h_{\rm ic}$, and (d) $h_{\rm fb}$.

16. The equivalent circuit common base h-parameters for a bipolar junction transistor are $h_{\rm ib} = 21.5$ Ω, $h_{\rm fb} = -0.95$, $h_{\rm rb} = 0.0031$, and $h_{\rm ob} = 4.7 \times 10^{-7}$ 1/Ω. What are this transistor's four common emitter h-parameters?

17. The h-parameters in the transistor shown are $h_{\rm ie} = 2200$ Ω, $h_{\rm re} = 3.6 \times 10^{-4}$, $h_{\rm fe} = 55$, and $h_{\rm oe} = 12.5$ μS. (a) What load resistance is needed for a signal output voltage of 0.1 V with an input signal current of 0.5 μA? (b) What input voltage is required? (c) What is the voltage gain (V_L/V_S)?

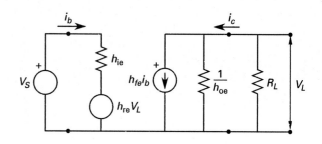

18. An AC-coupled FET amplifier circuit is shown.

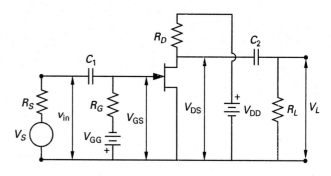

Assume the drain bias voltage is $V_{\rm DD} = 20$ V. If I_D at $V_{\rm DS} = 0$ V is 10 mA, draw the load line.

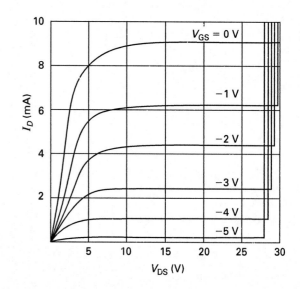

19. A FET with the characteristics shown is to be used in the circuit in Prob. 18. $R_D = 1.5$ kΩ, $V_{DD} = 20$ V, and $V_{GSQ} = -2$ V. Draw the DC load line and locate the Q-point.

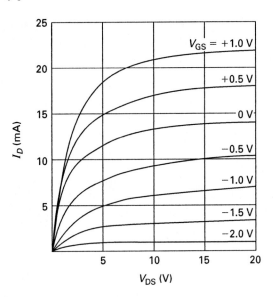

20. The manufacturer of a MOSFET has provided the following information: $V_{DS_{max}} = 20$ V, $V_{GS_{min}} = -8$ V, $I_{D_{max}} = 25$ mA, and $P_{max} = 330$ mW. The FET is to be used in the amplifier circuit shown, with $R_D = 800$ Ω and $R_L = 800$ Ω. Draw the DC load line and a Q-point to ensure linear amplification. Use the characteristic curves from Prob. 19.

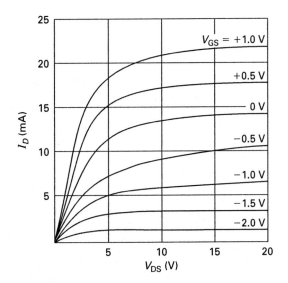

SOLUTIONS

1. In intrinsic semiconductors, $n = p$.

Thus,

$$n_i^2 = np = n^2$$

$$n_i = \boxed{2.5 \times 10^{13} \text{ cm}^{-3}}$$

The answer is (D).

2. "Acceptor" and "donor" describe types of impurities. Phosphorus is from group VA on the periodic table, with five valence electrons. Thus, it contributes a free electron. The semiconductor would be n-type.

The answer is (C).

3. Space-charge neutrality gives

$$N_A + n = N_D + p$$

Since $N_A = 0$ and $n \gg p$,

$$n \approx N_D = \boxed{10^{15} \text{ cm}^{-3}}$$

The answer is (D).

4. From Prob. 3, $N_D \approx n$. The law of mass action becomes

$$n_i^2 = np = N_D p$$

$$p = \frac{n_i^2}{N_D} = \frac{\left(1.6 \times 10^{10} \text{ cm}^{-3}\right)^2}{10^{15} \text{ cm}^{-3}}$$

$$= \boxed{2.56 \times 10^5 \text{ cm}^{-3} \quad (2.6 \times 10^5 \text{ cm}^{-3})}$$

The answer is (B).

5. Germanium has four valence electrons. The impurity has three valence electrons and thus adds a hole. The majority carriers will be holes.

The answer is (D).

6. Redraw the circuit as shown.

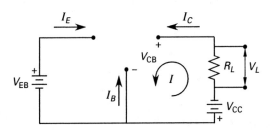

Kirchoff's voltage law (KVL) around the indicated loop gives

$$-V_{CC} - I_C R_L - V_{CB} = 0 \text{ V}$$

Because $I_C = 0$ A and $V_{CC} = -1$ V,

$$V_{CC} = V_{CB} = -1 \text{ V}$$

Redraw the current as shown.

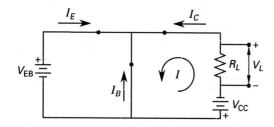

Using Ohm's law and the given information,

$$I_C = \frac{V_{CC}}{R_L} = \frac{-1 \text{ V}}{50 \text{ }\Omega} = -20 \text{ mA}$$

Plot V_{CC} and I_C, and connect them with a straight line. The result is normally plotted as shown.

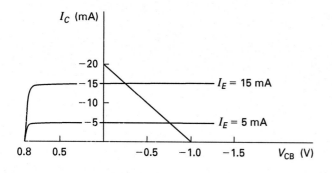

7. The diode current is

$$I = I_s \left(e^{\frac{V}{\eta V_T}} - 1 \right)$$

A reverse saturation current of 10^{-9} A is typical for silicon, $\eta = 2$, and V_T at room temperature is

$$V_T = \frac{\kappa T}{q} = \frac{\left(1.381 \times 10^{-23} \frac{\text{J}}{\text{K}} \right) (300\text{K})}{1.6 \times 10^{-19} \text{ C}}$$

$$= 0.026 \text{ V}$$

Substituting the known and calculated values gives the diode current.

$$I = \left(10^{-9} \text{ A} \right) \left(e^{\frac{0.75 \text{ V}}{(2)(0.026 \text{ V})}} - 1 \right)$$

$$= \boxed{1.84 \times 10^{-3} \text{ A} \quad (2.0 \times 10^{-3} \text{ A})}$$

The answer is (A).

8. The energy gap for GaP is 2.26 eV. The wavelength is

$$\lambda = \frac{hc}{E_G}$$

$$= \frac{\left(6.626 \times 10^{-34} \text{ J·s} \right) \left(3.00 \times 10^8 \frac{\text{m}}{\text{s}} \right)}{(2.26 \text{ eV}) \left(\frac{1.602 \times 10^{-19} \text{ J}}{1 \text{ eV}} \right)}$$

$$= \boxed{5.49 \times 10^{-7} \text{ m}}$$

The wavelength is in the visible spectrum.
The answer is (B).

9. (a)

$$I_s = \frac{I}{e^{\frac{qV}{\eta \kappa T}} - 1}$$

For a germanium diode, $\eta \approx 1$. At 20°C, $T_K = T_C + 273 = 293$K.

$$I_s = \frac{70 \text{ }\mu\text{A}}{e^{\left(\frac{(1.6 \times 10^{-19} \text{ C})(-1.5 \text{ V})}{(1) \left(\frac{1.38 \times 10^{-23} \text{ J}}{\text{K}} \right) (293\text{K})} \right)} - 1}$$

$$\approx \boxed{\frac{70 \text{ }\mu\text{A}}{-1} \quad (-70 \text{ }\mu\text{A})}$$

The reverse saturation current is constant for reverse voltages up to the breakdown voltage.

(b) $\quad I \approx I_s \left(e^{\frac{40V}{\eta}} - 1 \right)$

$$= (70 \times 10^{-6} \text{ A}) \left(e^{\frac{(40 \frac{1}{V})(0.2 \text{ V})}{1}} - 1 \right)$$

$$= \boxed{0.209 \text{ A}}$$

(c) $\dfrac{I_{s,40°C}}{I_{s,20°C}} = (2)^{\frac{40°C - 20°C}{10°C}}$

$$(2)^2 = 4$$

$$I_{s,40°C} = 4 I_{s,20°C} = (4)(70 \text{ }\mu\text{A})$$

$$= 280 \text{ }\mu\text{A}$$

$$I = I_s \left(e^{\frac{qV}{\eta \kappa T}} - 1 \right)$$

$$= (280 \times 10^{-6} \text{ A})$$

$$\times \left(e^{\left(\frac{(1.6 \times 10^{-19} \text{ C})(0.2 \text{ V})}{(1) \left(\frac{1.38 \times 10^{-23} \text{ J}}{\text{K}} \right) (313\text{K})} \right)} - 1 \right)$$

$$= \boxed{0.4617 \text{ A}}$$

10. Because $20 > 5\sqrt{2}$, the applied voltage is never negative. Therefore, the diode does not rectify the applied voltage and serves no purpose. Model the circuit as an ideal diode with a $100\,\Omega$ resistance. (The forward resistance could also be used.)

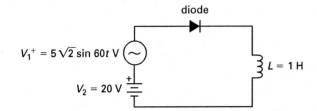

$$X_L = \omega L = \left(60\,\frac{\text{rad}}{\text{s}}\right)(1\text{ H}) = 60\,\Omega$$

$$Z = \sqrt{R^2 + X_L^2} = \sqrt{(100\,\Omega)^2 + (60\,\Omega)^2}$$

$$= 116.6\,\Omega$$

$$\phi = \arctan\frac{X_L}{R} = \arctan\frac{60\,\Omega}{100\,\Omega} = 30.96°$$

$$\mathbf{I} = \frac{\mathbf{V}}{\mathbf{Z}} = \frac{20 + 5\sqrt{2}\text{ V}\angle 0°}{116.6\,\Omega\angle 30.96°}$$

The DC voltage across the inductor is zero. Therefore, the 20 V bias is not used.

$$\mathbf{V}_L = \mathbf{IZ}_L = \left(\frac{5\sqrt{2}\text{ V}\angle 0°}{116.6\,\Omega\angle 30.96°}\right)(60\,\Omega\angle 90°)$$

$$= \boxed{3.639\text{ V}\angle 59.04°}$$

11. The saturation current doubles for every 10°C.

(a)
$$\frac{I_{s2}}{I_{s1}} = (2)^{\frac{\Delta T}{10°\text{C}}} = (2)^{\frac{100°\text{C}-25°\text{C}}{10°\text{C}}}$$

$$= (2)^{7.5} = 181$$

$$I_{s2} = (181)(100\ \mu\text{A})$$

$$= \boxed{18{,}100\ \mu\text{A}\quad(18.1\text{ mA})}$$

(b) At 0°C,

$$I_{s2} = (2)^{\frac{0°\text{C}-25°\text{C}}{10°\text{C}}}I_{s1} = (0.177)(100\ \mu\text{A})$$

$$= \boxed{17.7\ \mu\text{A}}$$

(c)
$$I_{-0.5} \approx I_s\left(e^{40V} - 1\right)$$

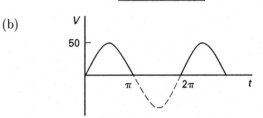

$$= (100\ \mu\text{A})\left(e^{\left(40\,\frac{1}{\text{V}}\right)(-0.5\text{ V})} - 1\right)$$

$$= \boxed{-100\ \mu\text{A}}$$

12. Clipping of the voltage wave will occur.

(a)
$$I = \frac{100\sin\omega t\text{ V}}{5000\,\Omega + 5000\,\Omega}$$

$$= \boxed{10^{-2}\sin\omega t\text{ A}}$$

(b)

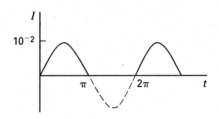

13.
$$I_C = \alpha I_E - I_{\text{CBO}}$$

$$= \alpha\left(\frac{V_{\text{BB}} - V_{\text{BE}}}{R_E}\right) - I_{\text{CBO}}$$

$$= (0.99)\left(\frac{4\text{ V} - 0.6\text{ V}}{1000\,\Omega}\right) - 10^{-6}\text{ A}$$

$$= \boxed{0.003365\text{ A}}$$

Alternate Solution

Writing KVL around the base loop,

$$V_{\text{BB}} - V_{\text{BE}} - (I_C + I_B)R_E = 0$$

$$\frac{I_C}{I_B} = \frac{\alpha}{1-\alpha} = \frac{0.99}{1-0.99} = 99$$

$$4\text{ V} - V_{\text{BE}} - \left(I_C + \frac{I_C}{99}\right)1000\,\Omega = 0$$

$$4\text{ V} - 0.6\text{ V} - (1010\,\Omega)I_C = 0$$

$$I_C = \boxed{3.366 \times 10^{-3}\text{ A}\quad(3.366\text{ mA})}$$

14. R_1 and R_2 form a voltage divider.

$$V_B = \left(\frac{R_2}{R_1 + R_2}\right) V_{CC}$$

$$= \left(\frac{60 \text{ k}\Omega}{30 \text{ k}\Omega + 60 \text{ k}\Omega}\right)(24 \text{ V})$$

$$= 16 \text{ V}$$

For the purpose of the signal, R_1 and R_2 both connect to ground and are in parallel.

$$R_{\text{in}} = \frac{R_1 R_2}{R_1 + R_2} = \frac{(30 \text{ k}\Omega)(60 \text{ k}\Omega)}{30 \text{ k}\Omega + 60 \text{ k}\Omega}$$

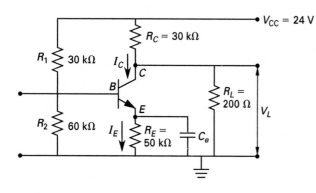

$$\beta = \frac{I_C}{I_B} = \frac{\alpha}{1 - \alpha} = \frac{0.95}{1 - 0.95} = 19$$

So, $I_B = I_C/19$.

Writing KVL around the base circuit,

$$V_B - V_{BE} - (I_C + I_B)R_E - I_B R_{\text{in}} = 0 \text{ V}$$

$$16 \text{ V} - 0.6 \text{ V} - \left(I_C + \frac{I_C}{19}\right)(50{,}000 \ \Omega)$$

$$- \left(\frac{I_C}{19}\right)(20{,}000 \ \Omega) = 0 \text{ V}$$

$$I_C = 2.869 \times 10^{-4} \text{ A} \ (0.2869 \text{ mA})$$

I_{CO} is given, so it should be used even though its effect is small.

$$I_C' = I_C + I_{CO} = 2.869 \times 10^{-4} \text{ A} + 10^{-6} \text{ A}$$

$$= \boxed{2.879 \times 10^{-4} \text{ A} \quad (0.2879 \text{ mA})}$$

The exact solutions, based on node-voltage analysis, are

$$V_B = 15.70 \text{ V}$$

$$I_B = 0.0151 \text{ mA}$$

$$I_C = 0.2869 \text{ mA}$$

15. (a) $h_{ib} = \dfrac{h_{ie}}{1 + h_{fe}} = \dfrac{500 \ \Omega}{1 + 100}$

$$= \boxed{4.95 \ \Omega}$$

(b) $h_{fc} = -1 - h_{fe} = -1 - 100$

$$= \boxed{-101}$$

(c) $h_{ic} = h_{ie}$

$$= \boxed{500 \ \Omega}$$

(d) $h_{fb} = \dfrac{-h_{fe}}{1 + h_{fe}} = \dfrac{-100}{1 + 100}$

$$= \boxed{-0.99}$$

16. $h_{ie} = \dfrac{h_{ib}}{1 + h_{fb}} = \dfrac{21.5 \ \Omega}{1 + (-0.95)}$

$$= \boxed{430 \ \Omega}$$

$$h_{re} = \frac{h_{ib} h_{ob}}{1 + h_{fb}} - h_{rb}$$

$$= \frac{(21.5 \ \Omega)\left(4.7 \times 10^{-7} \ \dfrac{1}{\Omega}\right)}{1 + (-0.95)} - 0.0031$$

$$= \boxed{-0.00290}$$

$$h_{fe} = \frac{-h_{fb}}{1 + h_{fb}} = \frac{-(-0.95)}{1 + (-0.95)}$$

$$= \boxed{19}$$

$$h_{oe} = \frac{h_{ob}}{1 + h_{fb}} = \frac{4.7 \times 10^{-7} \ \dfrac{1}{\Omega}}{1 + (-0.95)}$$

$$= \boxed{9.4 \times 10^{-6} \ \frac{1}{\Omega}}$$

17.

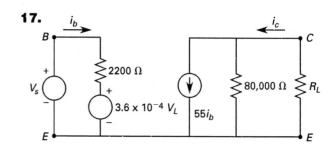

(a)
$$-i_c R_L = V_L$$

$$i_b = 0.5 \ \mu\text{A}$$

$$(55i_b)\left(\frac{80{,}000 R_L}{R_L + 80{,}000}\right) = V_L = 0.1 \ \text{V}$$

$$80{,}000 R_L = (R_L + 80{,}000)$$

$$\times \left(\frac{0.1}{(55)(0.5 \times 10^{-6} \ \text{A})}\right)$$

$$= 3636.36 R_L + 290.91 \times 10^6$$

$$R_L = \boxed{3809 \ \Omega}$$

(b)
$$V_s = i_b(2200 \ \Omega) + (3.6 \times 10^{-4})V_L$$

$$= (0.5 \times 10^{-6} \ \text{A})(2200 \ \Omega)$$

$$+ (3.6 \times 10^{-4})(0.1 \ \text{V})$$

$$= \boxed{1.136 \times 10^{-3} \ \text{V} \ (1.136 \ \text{mV})}$$

(c)
$$A = \frac{V_L}{V_s} = \frac{0.1 \ \text{V}}{1.136 \times 10^{-3}}$$

$$= \boxed{88.03}$$

18.
$$P_1 \ (V_{DS} = 20 \ \text{V}, \ I_D = 0 \ \text{A})$$

$$P_2 \ (V_{DS} = 0 \ \text{V}, \ I_D = 10 \ \text{mA})$$

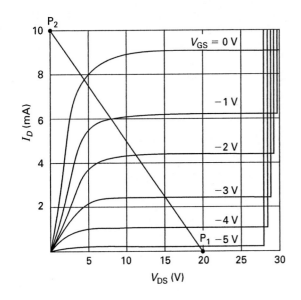

19. $P_1 \ (V_{DS} = 20 \ \text{V}, 0)$

$$P_2 \left(0, I_D = \frac{20 \ \text{V}}{1500 \ \Omega} = 0.01333 \ \text{A} \quad (13.33 \ \text{mA})\right)$$

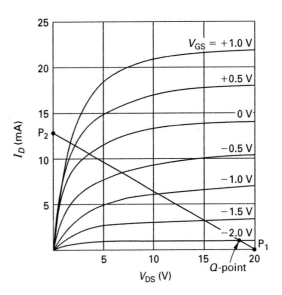

20. The highly negative V_{GS} curves are not shown and are, presumably, crowded together near the $I_D = 0$ A axis. Because of this, if the Q-point were in the highly negative region, positive swings in V_{GS} would accompany large increases in I_D, but negative swings in V_{GS} would decrease I_D very little. Therefore, the quiescent operating point needs to be near the $V_{GS} = 0$ V line. In this position, changes in I_D and V_{GS} are approximately proportional. Power dissipation is another consideration. It should be evaluated at the extreme points of the signal swing, not only at the Q-point.

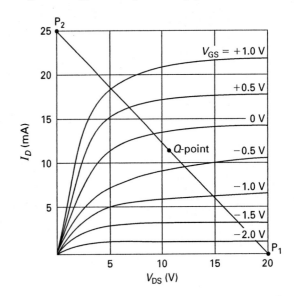

45 Amplifiers

PRACTICE PROBLEMS

1. An operational amplifier has a gain of 10^5. The frequencies at the half-power points are $f_L = 500$ Hz and $f_H = 1000$ Hz. What is the figure of merit?

(A) 1×10^6 rad/s
(B) 5×10^7 rad/s
(C) 3×10^8 rad/s
(D) 6×10^8 rad/s

2. During testing of an operational amplifier, a signal of equal magnitude is applied to the two terminals. That is, $v^+ = -v^-$. What gain is being measured?

(A) A_{cm}
(B) A_{dm}
(C) A_V
(D) G_P

3. An operational amplifier data sheet indicates a voltage gain of 10^5 and a power supply voltage of 12 V. What is the maximum voltage difference between the input terminals that will ensure linear operation?

(A) 60 μV
(B) 90 μV
(C) 120 μV
(D) 150 μV

4. Consider the circuit shown. The feedback resistance is 10 kΩ, and the input resistance is 100 Ω. If the op amp gain is 10^5, what is the gain of the circuit?

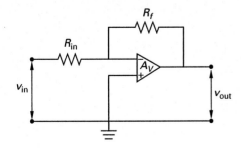

(A) -100
(B) -99.9
(C) $+100$
(D) 10^5

5. An operational amplifier has an input resistance of 10^5 Ω and a bandwidth of 1000 Hz, and is at a room temperature of 300K. If 10 dB above the noise is required for proper operation, what is the required signal power?

(A) 0.10 μV
(C) 2.0 μV
(C) 4.0 μV
(D) 16 μV

6. Consider the circuit shown. (a) If $R_f = 1$ MΩ and $R_{in} = 50$ Ω, what is the gain? (b) If $v_{in} = 1$ mV, what is v_{out}?

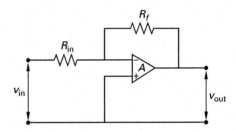

7. For the circuit shown, $v_1 = 10 \sin 200t$ and $v_2 = 15 \sin 200t$. What is v_{out}? The op amp is ideal and has infinite gain.

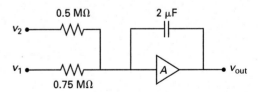

8. In the circuit shown, if v_{in} is $\sin 30t$, what is v_{out}? The op amp is ideal and has infinite gain.

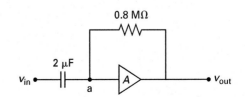

9. In the circuit shown, if $v_1 = v_2 = v_3 = 12$ V, what is v_{out}? The op amp is ideal and has infinite gain.

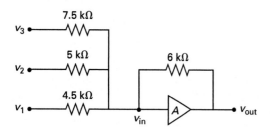

10. Given the following circuit, (a) what value of R_f will give an output of -8 V? (b) If the maximum of each input is 10 V and the output must not exceed -15 V, what value must be used for R_f? The op amp is ideal and has infinite gain.

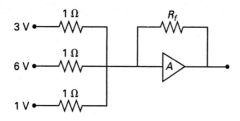

11. An integrator is used in a device for measuring displacement from a fixed location. The scale of the device is 1 V to 1 in. The switch is closed to set the output for initial displacement and then opened at $t = 0s$ to initiate measurement. If the displacement is initially 12 in d_0, what is the output voltage for an input of $8 \cos 10t$ V? The op amp is ideal and has infinite gain.

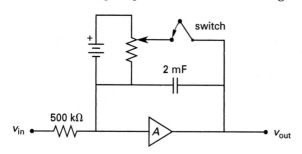

12. What differential equation is solved by the following circuit? The op amp is ideal and has infinite gain.

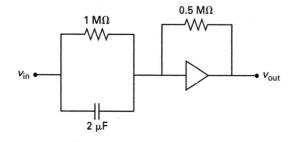

13. For the circuit shown, what is the differential equation that writes v_{out} in terms of v_{in}? The op amp is ideal and has infinite gain.

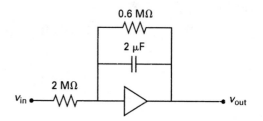

14. For the circuit shown, determine the output voltage as a function of the input voltage.

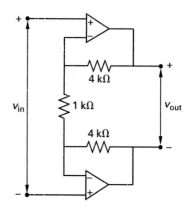

15. The op amp circuit shown is designed to provide a characteristic similar to a relay. When the output is negative and the input current increases, the output will go to the positive limit of $+12$ V when the input voltage exceeds $+1$ V. It will remain there until the input voltage becomes less than -1 V, at which time the output will go to the negative limit of -12 V. Determine an appropriate value for R.

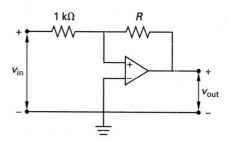

SOLUTIONS

1. The figure of merit is

$$F_m = A_{ref}(\text{BW}) = A_{ref}\left(2\pi f_H - 2\pi f_L\right)$$

$$= \left(10^5\right)\left(2\pi\right)\left(1000 \text{ Hz} - 500 \text{ Hz}\right)$$

$$= \boxed{3.14 \times 10^8 \text{ rad/s} \quad (3 \times 10^8 \text{ rad/s})}$$

The answer is (C).

2. The differential-mode voltage is

$$v_{dm} = v^+ - v^-$$

Thus, with $v^+ = -v^-$,

$$v_{dm} = v^+ + v^+ = 2v^+$$

The common-mode voltage is

$$v_{cm} = \tfrac{1}{2}\left(v^+ + v^-\right) = \tfrac{1}{2}\left(v^+ + \left(-v^-\right)\right) = 0 \text{ V}$$

Therefore, only the difference is being amplified, and the gain so determined is A_{dm}.

The answer is (B).

3. The maximum voltage difference is

$$\left|v^+ - v^-\right| < \frac{V_{DC} - 3 \text{ V}}{A_V}$$

$$|\Delta V| < \frac{12 \text{ V} - 3 \text{ V}}{10^5} = \frac{9 \text{ V}}{10^5}$$

$$= \boxed{90 \times 10^{-6} \text{ V} \quad (90 \text{ } \mu\text{V})}$$

The answer is (B).

4. Consider the circuit shown, assuming the indicated current directions at the negative terminal.

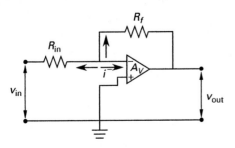

Writing Kirchoff's current law (KCL) at the inverting terminal node gives

$$\frac{V^- - v_{in}}{R_{in}} + \frac{V^- - v_{out}}{R_f} + i^- = 0 \text{ A}$$

Applying the ideal op amp assumptions means $i^- = 0$ A and $v^- = 0$ V. Substituting and solving for the gain gives

$$\frac{0 - v_{in}}{R_{in}} + \frac{0 - v_{out}}{R_f} + 0 = 0 \text{ A}$$

$$\frac{v_{out}}{v_{in}} = \frac{-R_f}{R_{in}} = A_{V,ideal}$$

$$= \frac{-10 \times 10^3 \text{ } \Omega}{100 \text{ } \Omega} = -100$$

But this op amp is real and has a finite gain. The constraint equation (that is, the fundamental op amp equation) rearranged is

$$V^- - V^+ = \Delta = \frac{-v_{out}}{A_V}$$

Since V^+ is at 0 V,

$$V^- = \frac{-v_{out}}{A_V}$$

Substituting into the original equation gives

$$\frac{\dfrac{-v_{out}}{A_V} - v_{in}}{R_{in}} + \frac{\dfrac{-v_{out}}{A_V} - v_{out}}{R_f} + 0 = 0$$

The input current is still small enough to be ignored. Rearranging gives

$$\frac{v_{out}}{v_{in}} = -A_V\left(\frac{R_f}{R_f + R_{in} + A_V R_{in}}\right)$$

Note that when the gain is large, this expression reduces to the expression for gain in the ideal case.

Substituting the given values gives the following result.

$$\frac{v_{out}}{v_{in}} = \left(-10^5\right)\left(\frac{10 \times 10^3 \text{ } \Omega}{\begin{array}{c}10 \times 10^3 \text{ } \Omega + 100 \text{ } \Omega \\ + (10^5)(100 \text{ } \Omega)\end{array}}\right)$$

$$= \boxed{-99.9}$$

The answer is (B).

5. The signal power is found from the signal to noise ratio (SNR).

$$\text{SNR} = 10 \log \frac{P_s}{P_n}$$

$$\text{antilog} \frac{\text{SNR}}{10} = \frac{P_s}{P_n}$$

$$P_s = P_n \text{ antilog} \frac{\text{SNR}}{10}$$

The noise power is unknown. The noise power is given in terms of the Boltzmann constant k; temperature, T, in kelvins; and the bandwidth, BW, in hertz.

$$P_n = \frac{V_{\text{rms}}^2}{4R} = (\text{BW})kT$$

$$= (1000 \text{ Hz}) \left(1.3805 \times 10^{-23} \frac{\text{J}}{\text{K}} \right) (300\text{K})$$

$$= 4.14 \times 10^{-18} \text{ W}$$

Substituting,

$$P_s = \left(4.14 \times 10^{-18}\text{W} \right) \text{antilog} \frac{10}{10}$$

$$= 41.4 \times 10^{-18} \text{ W}$$

The voltage is found from

$$P_s = \frac{V_{\text{rms}}^2}{4R}$$

$$V_{\text{rms}} = \sqrt{4RP_s} = \sqrt{(4)(10^5 \text{ }\Omega)(41.4 \times 10^{-18} \text{ W})}$$

$$= \boxed{4.07 \times 10^{-6} \text{ V} \quad (4.0 \text{ } \mu\text{V})}$$

The answer is (C).

6. (a) The gain is

$$A = \frac{-R_f}{R_{\text{in}}} = \frac{-10^6 \text{ }\Omega}{50 \text{ }\Omega} = \boxed{-20,000}$$

(b) $v_{\text{out}} = Av_{\text{in}} = (-20,000)(0.001 \text{ V}) = \boxed{-20 \text{ V}}$

7. This is a summing circuit.

$$v_{\text{out}} = -\left(\frac{Z_f}{Z_1} \right) v_1 - \left(\frac{Z_f}{Z_2} \right) v_2$$

$$= -\left(\frac{1}{sC} \right) \left(\frac{v_1}{R_1} + \frac{v_2}{R_2} \right)$$

$$= \frac{-1}{(2 \times 10^{-6} \text{ F})(0.75 \times 10^6 \text{ }\Omega)} \int 10 \sin 200t \text{ } dt$$

$$- \frac{1}{(2 \times 10^{-6} \text{ F})(0.5 \times 10^6 \text{ }\Omega)} \int 15 \sin 200t \text{ } dt$$

$$= \left(\frac{2}{3} \right) \left(\frac{1}{200} \right) (10) \cos 200t$$

$$+ \left(\frac{1}{1} \right) \left(\frac{15}{200} \right) \cos 200t + C$$

$$= \boxed{\left(\frac{1}{30} \right) \cos 200t + \left(\frac{3}{40} \right) \cos 200t + C \text{ V}}$$

8. The circuit is a differentiator.

$$v_{\text{out}} = -RC \frac{dv_{\text{in}}}{dt}$$

$$= -(0.8 \times 10^6 \text{ }\Omega)(2 \times 10^{-6} \text{ F}) \left(\frac{d(\sin 30t)}{dt} \right)$$

$$= \boxed{-48 \cos 30t \text{ V}}$$

9. This is a summing circuit.

$$v_{\text{out}} = \left(\frac{-6 \text{ k}\Omega}{4.5 \text{ k}\Omega} \right) v_1 - \left(\frac{6 \text{ k}\Omega}{5 \text{ k}\Omega} \right) v_2 - \left(\frac{6 \text{ k}\Omega}{7.5 \text{ k}\Omega} \right) v_3$$

$$= (-12 \text{ V}) \left(\frac{6 \text{ k}\Omega}{4.5 \text{ k}\Omega} + \frac{6 \text{ k}\Omega}{5 \text{ k}\Omega} + \frac{6 \text{ k}\Omega}{7.5 \text{ k}\Omega} \right)$$

$$= \boxed{-40 \text{ V}}$$

10. (a) This is a summing circuit.

$$v_{\text{out}} = - \left(\begin{array}{c} \left(\frac{R_f}{1 \text{ }\Omega} \right) (1 \text{ V}) + \left(\frac{R_f}{1 \text{ }\Omega} \right) (6 \text{ V}) \\ + \left(\frac{R_f}{1 \text{ }\Omega} \right) (3 \text{ V}) \end{array} \right)$$

$$= -8 \text{ V}$$

$$R_f = \boxed{0.8 \text{ }\Omega}$$

(b) $v_{\text{out}} = - \left(\begin{array}{c} \left(\dfrac{R_f}{1\ \Omega}\right)(10\text{ V}) + \left(\dfrac{R_f}{1\ \Omega}\right)(10\text{ V}) \\[2ex] + \left(\dfrac{R_f}{1\ \Omega}\right)(10\text{ V}) \end{array} \right)$

$= -15\text{ V}$

$R_f = \boxed{0.5\ \Omega}$

12. $\dfrac{v_{\text{out}}}{v_{\text{in}}} = -\dfrac{Z_f}{Z_{\text{in}}} = \dfrac{-500{,}000\ \Omega}{\dfrac{(10^6\ \Omega)\left(\dfrac{1}{s(2\times 10^{-6})}\ \Omega\right)}{10^6\ \Omega + \dfrac{1}{s(2\times 10^{-6})}\ \Omega}}$

$= -\dfrac{\frac{1}{2}}{\dfrac{1}{2s+1}} = -\left(s + \tfrac{1}{2}\right)$

$v_{\text{out}} = -\left(s + \tfrac{1}{2}\right) v_{\text{in}}$

$= \boxed{-\dfrac{dv_{\text{in}}}{dt} - \dfrac{v_{\text{in}}}{2}}$

11. $v_{\text{out}} = -\dfrac{1}{RC}\displaystyle\int v_{\text{in}}\, dt$

$= -\left(\dfrac{1}{(500{,}000\ \Omega) \times (2\times 10^{-3}\text{ F})}\right)\displaystyle\int 8\cos 10t\, dt$

$= (-8\times 10^{-4})\sin 10t + V_0$

At $T = 0$, $d_o = 12$ in.

13. $\dfrac{v_{\text{out}}}{v_{\text{in}}} = -\dfrac{Z_f}{Z_{\text{in}}}$

$= -\dfrac{\dfrac{(600{,}000\ \Omega)\left(\dfrac{1}{s(2\times 10^{-6})}\ \Omega\right)}{600{,}000\ \Omega + \left(\dfrac{1}{s(2\times 10^{-6})}\ \Omega\right)}}{2{,}000{,}000\ \Omega}$

$= -\dfrac{0.3}{1.2s+1}$

$v_{\text{out}}(1.2s + 1) = -0.3v_{\text{in}}$

$(1.2)\left(\dfrac{dv_{\text{out}}}{dt}\right) + v_{\text{out}} = -0.3v_{\text{in}}$

$\boxed{(1.2)\left(\dfrac{dv_{\text{out}}}{dt}\right) + v_{\text{out}} + 0.3v_{\text{in}} = 0}$

$V_0 = (12\text{ in})\left(\dfrac{1\text{ V}}{1\text{ in}}\right) = 12\text{ V}$

$v_{\text{out}} = \boxed{(-8\times 10^{-4})\sin 10t + 12\text{ V}}$

14.

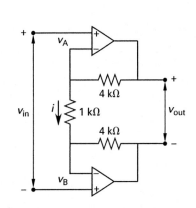

$$v_A = 0 \text{ V}$$

$$v_B = 0 \text{ V}$$

$$v_{in} = i(1 \text{ k}\Omega)$$

$$i = \frac{v_{in}}{1 \text{ k}\Omega}$$

$$v_{out} = (4 \text{ k}\Omega + 1 \text{ k}\Omega + 4 \text{ k}\Omega)i$$

$$= \left(\frac{9 \text{ k}\Omega}{1 \text{ k}\Omega}\right) v_{in}$$

$$= \boxed{9v_{in}}$$

15. Except while slewing from one saturation level to the other, the op amp is not operating in its linear region.

The current flowing in the resistors is

$$i_{in} = \frac{v_{in} - v_{out}}{1 \text{ k}\Omega + R}$$

The voltage of the positive terminal is

$$v_{in} - i_{in}(1 \text{ k}\Omega) = v_{in} - \frac{(1 \text{ k}\Omega)v_{in} - (1 \text{ k}\Omega)v_{out}}{1 \text{ k}\Omega + R}$$

$$v^+ = \frac{Rv_{in} + (1 \text{ k}\Omega)v_{out}}{R + 1 \text{ k}\Omega}$$

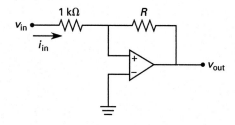

For v^+ positive, v_{out} is $+12$ V. v_{out} will reverse when v^+ becomes negative.

$$v_{in} = (-1 - \delta) \text{ V}$$

$$-R(1 \text{ V} + \delta) + (1 \text{ k}\Omega)(12 \text{ V}) < 0$$

$$-R + 12 \text{ k}\Omega - \delta R < 0$$

Let δ approach zero. Then,

$$R = \boxed{12 \text{ k}\Omega}$$

46 Pulse Circuits: Waveform Shaping and Logic

PRACTICE PROBLEMS

1. A germanium diode with an offset voltage of 0.2 V is used in a precision diode circuit. The power supply voltages are ± 15 V. The amplifier voltage gain is 10^7. What is the threshold voltage of the precision diode?

   (A) 20 nV

   (B) 100 nV

   (C) 1.5 μV

   (D) 9.0 μV

The following circuit is applicable to Probs. 2 through 5. The output is to be maintained at 24 V, while the input varies from 120 V to 126 V with a load current ranging from 0 A to 5 A. Assume ideal conditions, that is, the keep-alive current is zero and the zener resistance, R_Z, is negligible.

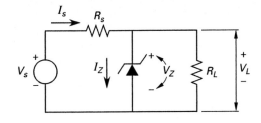

2. What is the required supply resistance to ensure proper operation of the zener regulator circuit?

   (A) 19 Ω

   (B) 20 Ω

   (C) 24 Ω

   (D) 25 Ω

3. What is the maximum zener current?

   (A) 4.1 A

   (B) 4.3 A

   (C) 5.1 A

   (D) 5.3 A

4. What is the minimum diode power rating that can be used?

   (A) 100 W

   (B) 150 W

   (C) 300 W

   (D) 600 W

5. What is the minimum power rating of the supply resistor?

   (A) 400 W

   (B) 450 W

   (C) 500 W

   (D) 550 W

The following circuit is applicable to Probs. 6 through 9. The circuit is to be operated at low frequencies. The transistor has the following h-parameter values: $h_{ie} = 4 \times 10^3$ Ω, $h_{fe} = 200$, $h_{oe} = 100 \times 10^{-6}$ S, and $h_{re} = 4 \times 10^3$.

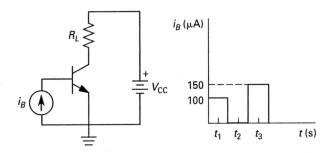

6. If the load resistance is 400 Ω, what collector supply voltage is required for the transistor to be on at time t_1?

   (A) 4 V

   (B) 8 V

   (C) 12 V

   (D) 16 V

7. If the collector supply voltage is 12 V and the load resistor is 400 Ω, what is the condition of the transistor at time t_2?

   (A) active

   (B) off

   (C) on

   (D) saturated

8. The selected collector supply voltage is 12 V. The transistor must be on for pulses of 150 μA or greater only. What is the required value of the load resistance?

   (A) 200 Ω

   (B) 300 Ω

   (C) 400 Ω

   (D) 600 Ω

9. For the circuit parameters in Prob. 8, what is the actual saturation current for a silicon transistor?

(A) 20.0 mA

(B) 25.5 mA

(C) 29.5 mA

(D) 30.0 mA

10. Consider the inverter, or NOT gate, circuit shown.

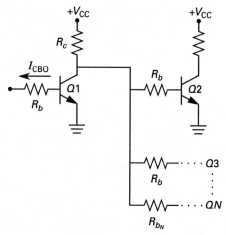

Parameters taken from a manufacturer's data sheet indicate that $V_{CC} = 5.0$ V, $I_{B,\text{sat}} > 0.5$ mA, $V_{BE,\text{sat}} < 1.5$ V, $R_c = 500$ Ω, and $R_b = 1000$ Ω. By how much does a reverse-saturation current of 1 mA lower the fan-out of $Q1$?

(A) 2

(B) 4

(C) 10

(D) 12

11. For the transistor circuit shown, determine (a) the necessary base current for this gate to sink five identical gates connected to its output, and (b) the minimum size of R_b to allow this gate to source five identical gates.

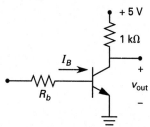

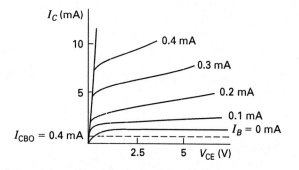

12. A particular logic gate has the following parameters: $t_d = 5$ ns, $t_r = 50$ ns, $t_s = 30$ ns, and $t_f = 30$ ns. Determine the maximum frequency at which this gate can operate.

13. Use an operational amplifier to convert a rectangular wave voltage of ± 5 V and a period of 10 ms to a triangular wave with a peak output of ± 5 V with the same period.

14. For the circuit shown, determine the transfer characteristic of the output voltage versus the input voltage. The diodes are ideal.

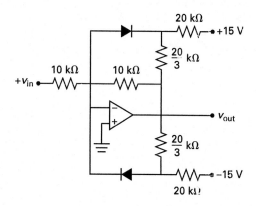

15. Determine the power requirement for the DTL circuit shown for a fan-out of five.

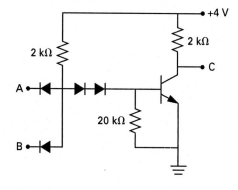

SOLUTIONS

1. The threshold voltage is reduced by an amount equal to the second term in Eq. 46.5, which is the transfer equation for a precision diode. The forward threshold voltage drop is V_F, and A_V is the voltage gain.

$$V_{\text{out}} = V_{\text{in}} - \frac{V_F}{A_V}$$

Substituting the given information gives

$$\frac{V_F}{A_V} = \frac{0.2 \text{ V}}{10^7} = \boxed{20 \times 10^{-9} \text{ V} \quad (20 \text{ nV})}$$

The answer is (A).

2. For a zener regulator circuit, the supply resistance is given by Eq. 46.8. V_{ZM} is the voltage zener at maximum rated current. $I_{L,\text{max}}$ is the maximum load current.

$$R_s = \frac{V_{\text{in,min}} - V_{\text{ZM}}}{I_{L,\text{max}}}$$

$$= \frac{120 \text{ V} - 24 \text{ V}}{5 \text{ A}}$$

$$= \boxed{19.2 \ \Omega \quad (19 \ \Omega)}$$

The answer is (A).

3. The maximum zener current is determined from Eq. 46.9.

$$I_{Z,\text{max}} = \left(\frac{V_{\text{in,max}} - V_{\text{ZM}}}{V_{\text{in,min}} - V_{\text{ZM}}}\right) I_{L,\text{max}}$$

$$= \left(\frac{126 \text{ V} - 24 \text{ V}}{120 \text{ V} - 24 \text{ V}}\right)(5 \text{ A})$$

$$= \boxed{5.3 \text{ A}}$$

The answer is (D).

4. The diode power is given by Eq. 46.10.

$$P_D = I_{Z,\text{max}} V_{\text{ZM}} = (5.3 \text{ A})(24 \text{ V})$$

$$= 127.2W$$

The minimum diode power rating that can be used is 150 W.

The answer is (B).

5. The minimum power rating of the supply resistor is given by Eq. 46.11.

$$P_{R_s} = I_{Z,\text{max}}^2 R_s$$

$$= (5.3 \text{ A})^2 (19.2 \ \Omega)$$

$$= \boxed{539.3 \text{ W} \quad (550 \text{ W})}$$

The answer is (D).

6. At low frequencies h_{fe} is approximately equal to h_{FE}, so the small-signal parameter given can be used. The base current, I_B, necessary for saturation is given by Eq. 46.25. V_{CC} is the collector supply voltage. R_L is the load resistance.

$$I_B \geq \frac{V_{\text{CC}}}{h_{\text{FE}} R_L}$$

For a load resistance of 400 Ω, the collector supply voltage is

$$V_{\text{CC}} \leq I_{B_{t_1}} h_{\text{FE}} R_L$$

$$\leq (100 \times 10^{-6} \text{ A})(200)(400 \ \Omega)$$

$$\leq \boxed{8 \text{ V}}$$

The answer is (B).

7. If the base current is zero, the transistor is off.

The answer is (B).

8. The minimum base current, which will be 150×10^{-6} A to ensure that the transistor is not on at t_1, is given by Eq. 46.25.

$$I_B \geq \frac{V_{\text{CC}}}{h_{\text{FE}} R_L} \approx \frac{V_{\text{CC}}}{h_{\text{fe}} R_L}$$

Rearranging gives

$$R_L \geq \frac{V_{\text{CC}}}{h_{\text{fe}} I_B}$$

$$\geq \frac{12 \text{ V}}{(200)(150 \times 10^{-6} \text{ A})}$$

$$\geq \boxed{400 \ \Omega}$$

The answer is (C).

9. The actual saturation current is given by Eq. 46.24. $V_{\text{CE,sat}}$ is the collector emitter voltage at saturation.

$$I_{C,\text{sat}} = \frac{V_{\text{CC}} - V_{\text{CE,sat}}}{R_L}$$

$$= \frac{12 \text{ V} - 0.2 \text{ V}}{400 \ \Omega}$$

$$= \boxed{29.5 \times 10^{-3} \text{ A} \quad (29.5 \text{ mA})}$$

Note that if the collector-emitter voltage drop is ignored, the saturation current will be 30×10^{-3} A. Thus, except for exacting calculations, the ON transistor can be modeled as a short circuit.

The answer is (C).

10. A transistor inverter is a common-emitter configured transistor as shown.

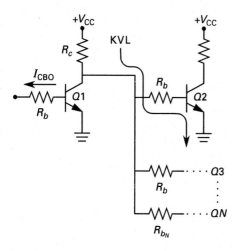

Initially ignoring the reverse saturation current, I_{CBO}, means $Q2$ can be modeled as an open circuit. Then, writing Kirchoff's voltage law (KVL) from the reference ground at $+V_{\mathrm{CC}}$ (not shown) back to the reference ground (shown at $Q2$), which comprises a loop, gives

$$V_{\mathrm{CC}} - NI_{B,\min}R_c - I_{B,\min}R_b - V_{\mathrm{BE2}} = 0$$

The minimum base current is the current necessary to cause $Q2$ and other load transistors to operate in the saturation region. This is given as $I_{B,\mathrm{sat}} > 0.5$ mA. The base-emitter voltage is the voltage of the transistor while saturated, given as $V_{\mathrm{BE,sat}} < 1.5$ V. Substituting and solving for the fan-out gives

$$N = \frac{V_{\mathrm{CC}} - V_{\mathrm{BE,sat}} - I_{B,\min}R_b}{I_{B,\min}R_c}$$

$$= \frac{5\ \mathrm{V} - 1.5\ \mathrm{V} - (0.5 \times 10^{-3}\ \mathrm{A})(1000\ \Omega)}{(0.5 \times 10^{-3}\ \mathrm{A})(500\ \Omega)}$$

$$= 12$$

To account for I_{CBO}, add the additional voltage drop $I_{\mathrm{CBO}}R_c$ to the original KVL equation, or use Eq. 46.33 directly.

$$V_{\mathrm{CC}} - NI_{B,\min}R_c - I_{\mathrm{CBO}}R_c - I_{B,\min}R_b - V_{\mathrm{BE2}} = 0$$

$$N + \frac{R_b}{R_c} = \frac{V_{\mathrm{CC}} - V_{\mathrm{BE2}} - I_{\mathrm{CBO}}R_c}{I_{B,\min}R_c}$$

Substituting and solving gives

$$N = 10$$

The reverse saturation current lowers the fan-out by

$$12 - 10 = \boxed{2}$$

The answer is (A).

11. The load line with no loads has $I_C \approx 4.9$ mA, which requires I_B to be slightly more than 0.3 mA.

(a) To sink five additional loads, because $I_{\mathrm{CBO}} = 0.4$ mA, I_C must increase by $(5)(0.4\ \mathrm{mA}) = 2$ mA to approximately 6.9 mA. This will require $I_B \approx 0.4$ mA.

$$I_{B,\mathrm{low\ output}} = \boxed{0.4\ \mathrm{mA}}$$

(b) When this transistor is off, the five loads draw 0.4 mA each, while the bases have 0.7 V. The 0.4 mA sourcing transistor's reverse saturation current must also be accounted for, as shown.

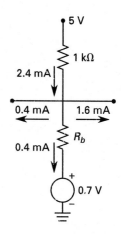

$$5\ \mathrm{V} = (2.4\ \mathrm{mA})(1\ \mathrm{k}\Omega) + (0.4\ \mathrm{mA})R_b + 0.7\ \mathrm{V}$$

$$R_b = \frac{5\ \mathrm{V} - 0.7\ \mathrm{V} - 2.4\ \mathrm{V}}{(0.4\ \mathrm{mA})\left(0.001\ \dfrac{\mathrm{A}}{\mathrm{mA}}\right)}$$

$$= \boxed{4750\ \Omega \quad (4.75\ \mathrm{k}\Omega)}$$

12.
$$\tau_{\min} = \tau_d + \tau_r + \tau_s + \tau_f = 115\ \mathrm{ns}$$

$$= 5\ \mathrm{ns} + 50\ \mathrm{ns} + 30\ \mathrm{ns} + 30\ \mathrm{ns}$$

$$f_{\max} = \frac{1}{\tau_{\min}} = \frac{10^9\ \dfrac{\mathrm{s}}{\mathrm{ns}}}{115\ \mathrm{ns}}$$

$$= 8.7 \times 10^6\ \mathrm{Hz} \quad \boxed{(8.7\ \mathrm{MHz})}$$

13. The desired conversion is

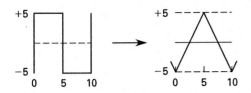

The required circuit is an integrator, shown in the following illustration.

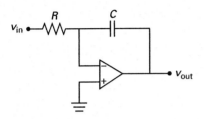

Use Kirchoff's current law (KCL) at the inverting terminal to get

$$\frac{0 - v_{in}}{R} + C\frac{d(0 - v_{out})}{dt} = 0$$

Rearranging and integrating gives

$$v_{out} = -\frac{1}{RC}\int v_{in} dt + \kappa$$

At half the period, 5 ms, the necessary initial conditions are the level of the voltage and the time frame.

$$v_{out} = -\frac{1}{RC}\int_{0\,s}^{0.005\,s} 5 dt + 5\text{ V} = -5\text{ V}$$

$$= -\frac{5\text{ V}}{RC}t\Big|_{0\,s}^{0.005\,s} + 5\text{ V} = -5\text{ V}$$

$$= -\frac{5\text{ V}}{RC}(0.005\text{ s}) = -10\text{ V}$$

Solve for the combination RC.

$$-\frac{5\text{ V}}{RC}(0.005\text{ s}) = -10\text{ V}$$

$$RC = -\frac{(5\text{ V})(0.005\text{ s})}{-10\text{ V}}$$

$$= \boxed{2.5 \times 10^{-3}\text{ s}}$$

Therefore, choosing an RC combination value equal to the calculated value provides the transient necessary to effect the desired conversion.

14. With both diodes open,

$$\frac{v_{out}}{v_{in}} = -1$$

With the upper diode conducting,

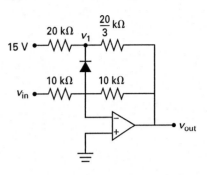

$$\frac{\left(\frac{20}{3}\text{ k}\Omega\right)(10\text{ k}\Omega)}{\left(\frac{20}{3}\text{ k}\Omega\right) + 10\text{ k}\Omega} = 4\text{ k}\Omega$$

$$\frac{15\text{ V}}{20\text{ k}\Omega} + \frac{v_{in}}{10\text{ k}\Omega} + \frac{v_{out}}{4\text{ k}\Omega} = 0$$

$$v_{out} = -0.4v_{in} - 3\text{ V}$$

For the lower diode conducting,

$$v_{out} = -0.4v_{in} + 3\text{ V}$$

For both diodes open,

$$v_1 = (15\text{ V})\left(\frac{20\text{ k}\Omega}{\frac{20}{3}\text{ k}\Omega + 20\text{ k}\Omega}\right)$$

$$+ v_{out}\left(\frac{20\text{ k}\Omega}{\frac{20}{3}\text{ k}\Omega + 20\text{ k}\Omega}\right)$$

$$= \frac{(15\text{ V})\left(\frac{20}{3}\text{ k}\Omega\right) + v_{out}(20\text{ k}\Omega)}{\frac{20}{3}\text{ k}\Omega + 20\text{ k}\Omega}$$

$$= \frac{15\text{ V} + 3v_{out}}{4} = \frac{15\text{ V} - 3v_{in}}{4}$$

Because the negative terminal is at virtual ground, for the upper diode to be off, $v_1 > 0$ V, so

$$15\text{ V} - 3v_{in} > 0\text{ V}$$

$$v_{in} < 5\text{ V}$$

For the lower diode,

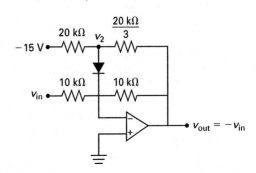

For both diodes open,

$$v_2 = (-15 \text{ V}) \left(\frac{\frac{20}{3} \text{ k}\Omega}{\frac{20}{3} \text{ k}\Omega + 20 \text{ k}\Omega} \right)$$

$$- v_{\text{in}} \left(\frac{\frac{20}{3} \text{ k}\Omega}{\frac{20}{3} \text{ k}\Omega + 20 \text{ k}\Omega} \right)$$

$$= \frac{(-15 \text{ V}) \left(\frac{20}{3} \text{ k}\Omega \right) - v_{\text{in}}(20 \text{ k}\Omega)}{\frac{20}{3} \text{ k}\Omega + 20 \text{ k}\Omega}$$

$$= \frac{-15 \text{ V} - 3v_{\text{in}}}{4} < 0 \text{ V}$$

$$v_{\text{in}} > -5 \text{ V}$$

For $-5 \text{ V} < v_{\text{out}} < 5 \text{ V}, v_{\text{out}} = -v_{\text{in}}$

$$v_{\text{in}} < -5 \text{ V}, v_{\text{out}} = -0.4v_{\text{in}} + 3 \text{ V}$$

$$v_{\text{in}} > 5 \text{ V}, v_{\text{out}} = -0.4v_{\text{in}} - 3 \text{ V}$$

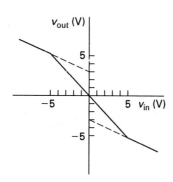

15. With at least one input low, the collector is high and no current flows in the collector circuit.

One or more low inputs of 0.2 V, when combined with a diode drop of 0.7 V, results in a current through the left resistor of

$$I = \frac{\Delta V}{R}$$

$$I = \frac{4 \text{ V} - 0.7 \text{ V} - 0.2 \text{ V}}{(2 \text{ k}\Omega) \left(1000 \frac{\Omega}{\text{k}\Omega} \right)}$$

$$= 1.55 \times 10^{-3} \text{ A} \quad (1.55 \text{ mA})$$

The power for high output is

$$P = IV$$

$$= (1.55 \text{ mA})(4 \text{ V})$$

$$= 6.2 \text{ mW}$$

With both inputs high,

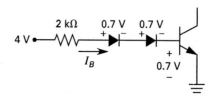

$$I = \frac{\Delta V}{R}$$

$$I_B = \frac{4 \text{ V} - (3)(0.7 \text{ V})}{(2 \text{ k}\Omega) \left(1000 \frac{\Omega}{\text{k}\Omega} \right)}$$

$$= 0.95 \times 10^{-3} \text{ A} \quad (0.95 \text{ mA})$$

This saturates the transistor, so

$$I = \frac{\Delta V}{R}$$

$$I_C = \frac{4 \text{ V} - 0.2 \text{ V}}{(2 \text{ k}\Omega) \left(1000 \frac{\Omega}{\text{k}\Omega} \right)}$$

$$= 1.9 \times 10^{-3} \text{ A} \quad (1.9 \text{ mA})$$

The power supply provides

$$P_{\text{supply}} = IV$$

$$(1.9 \text{ mA} + 0.95 \text{ mA})(4 \text{ V}) = 11.4 \text{ mW}$$

However, the transistor must absorb power from the loads, which must be calculated from a current of five times the input current for the high output. This excess collector current is

$$I_c = 5I = (5)(1.55 \text{ mA}) = 7.75 \text{ mA}$$

Assuming the transistor remains saturated, the excess transistor power can be as high as

$$P_{\text{excess}} = IV$$

$$= (7.75 \text{ mA})(0.2 \text{ V})$$

$$= 1.55 \text{ mW}$$

The maximum total power is

$$P_{\text{total}} = P_{\text{supply}} + P_{\text{excess}}$$

$$= 1.55 \text{ mW} + 11.4 \text{ mW}$$

$$= \boxed{12.95 \text{ mW}}$$

47 Computer Hardware Fundamentals

PRACTICE PROBLEMS

1. A disk has 1000 tracks with 10,000 bits per track. The disk diameter is 15.24 cm. The average seek time is 20 ms, and the rotational speed is 2000 rpm. (a) What is the areal density of the disk? (b) Estimate the average access time.

2. What length of nine-track tape with density 1600 Bpi is required to store 1 MB of data?

3. The entire contents of a full 20 MB disk are sent in an asynchronous transmission at a 2400 baud rate. What is the time required for transmission?

4. The instruction format shown uses one memory word per instruction.

```
11 10        54        0
 P |  CODE  | ADDRESS |
```

What is the maximum bit size of user memory?
- (A) 16 bits
- (B) 192 bits
- (C) 4096 bits
- (D) 65 536 bits

5. What is the term used to describe the time it takes to locate and read RAM data?
- (A) access time
- (B) buffering time
- (C) dynamic response
- (D) interface time

SOLUTIONS

1. (a) areal density

$$= \text{(no. bits per track)}$$
$$\times \text{(no. tracks per radial centimeter)}$$

$$= \left(10{,}000 \, \frac{\text{bits}}{\text{track}}\right)\left(\frac{1000 \text{ tracks}}{\frac{D}{2}}\right)$$

$$= \left(10{,}000 \, \frac{\text{bits}}{\text{track}}\right)\left(\frac{2000 \text{ tracks}}{15.24 \text{ cm}}\right)$$

$$= \boxed{1.31 \times 10^6 \text{ bits/radial cm}}$$

(b) access time = average latency + average seek time

The average latency can be approximated as one-half the time for the full revolution.

$$\left(\tfrac{1}{2} \text{ rev}\right)\left(\frac{1 \text{ min}}{2000 \text{ rev}}\right)\left(60 \, \frac{\text{sec}}{\text{min}}\right)\left(1000 \, \frac{\text{ms}}{\text{sec}}\right) = 15 \text{ ms}$$

average access time $= 15 \text{ ms} + 20 \text{ ms}$

$$= \boxed{35 \text{ ms}}$$

2. Because eight tracks are used for data recording, 1 byte of data is stored across the width of the tape. Therefore, 1600 bytes of data are stored per 2.54 cm of length. The length L required for 1 MB is

$$L = \frac{1{,}048{,}576 \text{ bytes}}{\left(\frac{1600 \text{ bytes}}{2.54 \text{ cm}}\right)\left(100 \frac{\text{cm}}{\text{m}}\right)} = \boxed{16.6 \text{ m}}$$

This neglects any interblock gaps.

3. Asynchronous, or start-stop, transmissions use a start and stop bit for each byte. Thus, there are 10 bits/byte. The time required is

$$t = \frac{(20 \text{ MB})\left(1{,}048{,}576 \, \frac{\text{bytes}}{\text{MB}}\right)\left(10 \, \frac{\text{bits}}{\text{byte}}\right)}{\left(2400 \, \frac{\text{bits}}{\text{sec}}\right)\left(60 \, \frac{\text{sec}}{\text{min}}\right)\left(60 \, \frac{\text{min}}{\text{hr}}\right)}$$

$$= \boxed{24.3 \text{ hr}}$$

4. The address field is 5 bits wide [4:0]. Each word is 12 bits [11:0]. Therefore, there are 2^4 addresses, each containing a 12-bit word.

$$(2^4 \text{ addresses}) \left(12 \, \frac{\text{bits}}{\text{address}} \right) = \boxed{192 \text{ bits}}$$

The answer is (B).

5. The time it takes to locate and read data stored in random access memory (RAM) is called access time.

The answer is (A).

48 Computer Software Fundamentals

PRACTICE PROBLEMS

1. Flowchart the following procedure: If A is greater than B and A is greater than C, then add B to C and go to PLACE1. If B is greater than A and B is greater than C, then add A to C and go to PLACE2. If B is greater than A but less than C, then subtract A from B and go to PLACE1.

2. Flowchart the following procedure: If A is greater than 10 and if A is less than 14, subtract X from A. If A is less than 10 or if A is greater than 14, exit the program.

3. Draw a flowchart representing an algorithm that finds the real roots of a second-degree equation with real coefficients. The coefficients are to be read from an external peripheral, and the results are to be printed.

4. Draw a flowchart representing the bubble sort algorithm. (Sort into ascending order.)

5. For the binary search procedure, derive the number of required comparisons in a list of n elements.

SOLUTIONS

1.

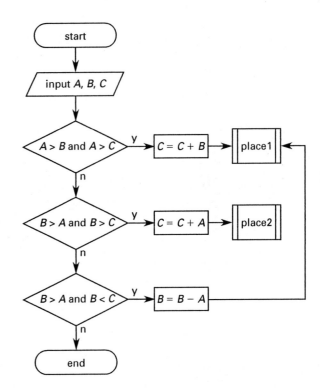

2.

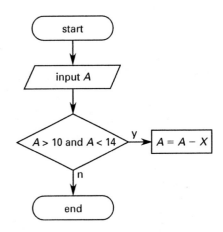

Computers

3.

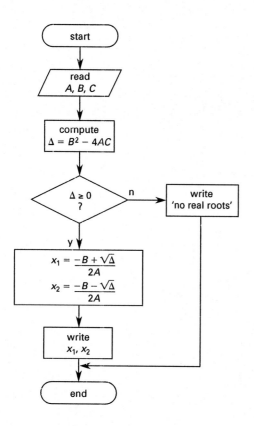

4.

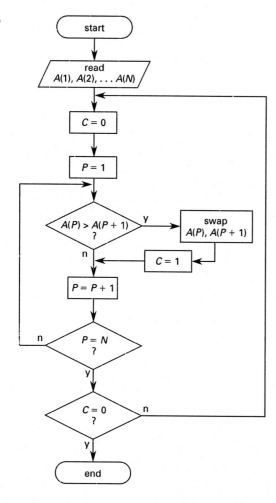

5. To simplify, assume that the number of elements in the list is some integer power of two.

$$n = 2^p$$

Assume that the element sought for is the first one in the list. This will produce a worst-case search. Let $a(n)$ be the number of comparisons needed to find the element in the list. Because the binary search procedure discards (disregards) half of the list after each comparison,

$$a(n) = 1 + a\left(\frac{n}{2}\right) \qquad \text{[Eq. I]}$$

But, $n = 2^p$.

$$a(n) = a2^p = 1 + a\left(\frac{2^p}{2}\right) = 1 + a2^{p-1} \qquad \text{[Eq. II]}$$

Equation I can be applied to the second term of Eq. II.

$$a2^{p-1} = 1 + a\left(\frac{2^{p-1}}{2}\right) = 1 + a2^{p-2} \qquad \text{[Eq. III]}$$

Combining Eqs. II and III,

$$a(n) = a2^p = 1 + 1 + a2^{p-2}$$

Equation I can continue to be applied to the last term until the list contains only two elements, after which the search becomes trivial. This results in a string of p ones.

$$a(n) = 1 + 1 + 1 + \cdots + 1 = p$$

An argument for $a(n) = p + 1$ can be made depending on how the search logic handles a miss with two elements in the list. If it merely selects the remaining element without comparing it, then $a(n) = p$. If a singular element is investigated, then $a(n) = p + 1$.

Now,

$$n = 2^p$$

$$\log_2 n = \log_2 2^p = p$$

But $p = a(n) = \log_2 n$.

Using logarithm identities,

$$\boxed{p = \log_2 n = \frac{\log_{10} n}{\log_{10} 2}}$$

49 Computer Architecture

PRACTICE PROBLEMS

1. Which of the following is NOT a type of information?

(A) character
(B) data
(C) logical
(D) numerical

2. What type of instruction set is used for a read/write (R/W) command?

(A) arithmetic
(B) logic
(C) memory
(D) control

3. What classification is used to describe a computer that handles instruction sets using a single high-level language instruction?

(A) CISC
(B) DISC
(C) PC
(D) RISC

4. A control line uses four bits of information. What is the maximum number of operations possible?

(A) 4 operations
(B) 8 operations
(C) 15 operations
(D) 16 operations

5. A register is accessed using double-word granularity. How many bits of information are being accessed at one time?

(A) 4 bits
(B) 8 bits
(C) 16 bits
(D) 32 bits

SOLUTIONS

1. "Data" is a general term for the quantities that input to a computer and is not a type of information.

The answer is (B).

2. A read/write (R/W) command involves the control and transfer of information.

The answer is (D).

3. The use of a single high-level language instruction is meant to simplify the programming process for the designer but involves more complex computer-level instructions. Thus, a computer that uses a single high-level language instruction is classified is as a complex instruction-set computer (CISC).

The answer is (A).

4. A 4-bit line can contain $2^4 = 16$ different signals.

The answer is (D).

5. Two bytes (16 bits) are considered to be a single word. Double-word granularity, then, is 32 bits.

The answer is (D).

50 Digital Logic

PRACTICE PROBLEMS

1. What is the state of a TTL cell at 4.0 V?

(A) 0

(B) indeterminate

(C) 1

(D) saturated

2. A computer accesses a register in 4-bit packets, sometimes called a *nibble*. What is the total number of values that can be represented in the nibble?

(A) 4

(B) 8

(C) 15

(D) 16

3. A computer data sheet indicates a 64K-byte RAM. How many bytes of memory are available?

(A) 8 bytes

(B) 1024 bytes

(C) 64,000 bytes

(D) 65,536 bytes

4. How many bits of information are contained within a 64K-byte RAM?

(A) 64 bits

(B) 8192 bits

(C) 512,000 bits

(D) 524,288 bits

5. If $x_1 = 0$, $x_2 = 1$, and $x_3 = 0$, what is the value of the following expression?

$$f(x_1, x_2, x_3) = \overline{x}_1 x_2 x_3 + x_1 \overline{x}_2 x_3 \cdot x_1 x_2 \overline{x}_3$$

6. A four-variable function has a minterm in row five. That minterm is given by which of the following expressions?

(A) $\overline{A}B\overline{C}$

(B) $A\overline{B}C$

(C) $\overline{A}B\overline{C}D$

(D) $A\overline{B}C\overline{D}$

The following truth table is applicable to Probs. 7 and 8.

A	B	C	$F(A,B,C)$
0	0	0	0
0	0	1	0
0	1	0	0
0	1	1	1
1	0	0	0
1	0	1	0
1	1	0	0
1	1	1	1

7. What is the canonical sum-of-products (SOP) expression for the function in the truth table shown?

(A) $\overline{A}\,\overline{B}\,\overline{C} + \overline{A}\,BC + \overline{A}B\overline{C}$

(B) $A\overline{B}\,\overline{C} + AB\,\overline{C} + AB\overline{C}$

(C) (A) and (B) combined

(D) $\overline{A}BC + ABC$

8. What maxterms are necessary to create the canonical product-of-sums (POS) expression for the function in the truth table shown?

(A) $M_0 M_1 M_2$

(B) $M_4 M_5 M_6$

(C) (A) and (B) combined

(D) $M_3 M_7$

The following truth table is applicable to Probs. 9 and 10.

A	B	$F(A,B)$
−5 V	−5 V	−5 V
−5 V	0	0
0	−5 V	0
0	0	−5 V

9. If positive logic is applied to the truth table, what logic operation is obtained?

(A) coincidence

(B) NOR

(C) XOR

(D) XNOR

10. If negative logic is applied to the truth table, what logic operation is obtained?

   (A) coincidence

   (B) NOR

   (C) OR

   (D) XOR

SOLUTIONS

1. The state of a cell is its binary value. For TTL, any voltage between 2.0 V and 5.0 V is state 1.

The answer is (C).

2. The total number of values in any n-tuple logic expression is given by Eq. 50.1.

$$N = 2^n = 2^4 = \boxed{16}$$

The answer is (D).

3. The K indicates $2^{10} = 1024$. Thus,

$$64\text{K} = (64)(2)^{10} = \boxed{65{,}536 \text{ bytes}}$$

The answer is (D).

4. One byte is 8 bits. Thus,

$$(65{,}536 \text{ bytes})\left(\frac{8 \text{ bits}}{1 \text{ byte}}\right) = \boxed{524{,}288 \text{ bits}}$$

The answer is (D).

5. The function is

$$f(x_1, x_2, x_3) = \overline{x}_1 x_2 x_3 + x_1 \overline{x}_2 x_3 \cdot x_1 x_2 \overline{x}_3$$

Substituting the given values and using the order of precedence rules gives

$$f(0,1,0) = (\overline{0} \cdot 1 \cdot 0) + (0 \cdot \overline{1} \cdot 0) \cdot (0 \cdot 1 \cdot \overline{0})$$

$$= (1 \cdot 1 \cdot 0) + (0 \cdot 0 \cdot 0) \cdot (0 \cdot 1 \cdot 1)$$

$$= 0 + 0 \cdot 0$$

$$= 0 + 0$$

$$= \boxed{0}$$

6. A function with four variables (that is, with $n = 4$) with a value of one in row five has a minterm in row five. Changing row five from decimal to binary gives

$$5_{10} = 0101_2$$

The minterm is deduced from its binary representation as

$$\boxed{\overline{A}B\overline{C}D = m_5}$$

The answer is (C).

7. The minterms are in rows three and seven—that is, where the function equals one. The canonical sum-of-product form is

$$F(A,B,C) = m_3 + m_7 = \boxed{\overline{A}BC + ABC}$$

The answer is (D).

8. The maxterms are in the rows where the function has a value of zero. Thus,

$$F(A,B,C) = \boxed{\prod M(0,1,2,4,5,6)}$$

The answer is (C).

9. If positive logic is applied, the high voltage has a value of one. Rewriting the truth table with -5 V as logic 0 and 0 V as logic 1 gives

A	B	$F(A,B)$
0	0	0
0	1	1
1	0	1
1	1	0

This is the exclusive OR (XOR) operation.

The answer is (C).

10. If negative logic is applied, the low voltage has a value of one. Rewriting the truth table with -5 V as logic 1 and 0 V as logic 0 gives

A	B	$F(A,B)$
1	1	1
1	0	0
0	1	0
0	0	1

Reversing the order to place the table in standard format gives

A	B	$F(A,B)$
0	0	1
0	1	0
1	0	0
1	1	1

This is the XNOR, or coincidence, logic.

The answer is (A).

Computers

PRACTICE PROBLEMS

1. A 3-tuple logic expression has maxterms M_1, M_2, and M_3. Which of these maxterms are adjacent?

(A) M_1 and M_2

(B) M_1 and M_3

(C) M_2 and M_3

(D) both (B) and (C)

2. What is the POS expression for the truth table shown?

(A) $\overline{C} \cdot B + \overline{C} \cdot A$

(B) $(\overline{C} + B) \cdot (\overline{C} + A)$

(C) $(A + B) \cdot (A + \overline{C}) \cdot (\overline{A} + \overline{C})$

(D) $(A + B) \cdot (\overline{C})$

3. What type of logic directly realizes the maxterm formulation of the truth table in Prob. 2?

(A) AND

(B) AND-OR

(C) NAND-NAND

(D) NOR-NOR

4. What is the minimal expression that can be obtained using the given Karnaugh map only?

(A) $ABCD + A\overline{B}C + \overline{A}BC + C\overline{D} + \overline{B}\,\overline{D}$

(B) $\overline{A}\,\overline{B}\,\overline{C}\,\overline{D} + A\overline{B}C + \overline{A}BC + C\overline{D} + \overline{B}\,\overline{D}$

(C) $\overline{A}\,\overline{B}C\overline{D} + \overline{A}B\overline{C} + A\overline{B}\,\overline{C} + \overline{C}D + BD$

(D) none of the above

5. The design of a coin-changer results in the following truth table. What is the minimal logic expression that results in the truth table output?

(A) $A + \overline{C}$

(B) $\overline{A} + \overline{C}$

(C) $B + \overline{C}$

(D) $\overline{B} + C$

6. What is the NOR-only expression for

$$F(A,B,C) = \prod M(0,1,2,5)$$

(A) $\overline{\overline{(A+C)} + \overline{(B+\overline{C})}}$

(B) $\overline{(A+B)\,(A+C)}\ \overline{(\overline{A}+B+C)}$

(C) $\overline{\overline{(A+B)} + \overline{(A+C)} + \overline{(\overline{A}+B+C)}}$

(D) $\overline{\overline{(A+B)} + \overline{(A+C)} + \overline{(\overline{A}+B+\overline{C})}}$

7. For the truth table given, obtain the Karnaugh map and Veitch diagram for the outputs D, E, and F. An X entry in the truth table indicates a don't care, so zero or one can be used for maximum convenience.

inputs			outputs		
A	B	C	D	E	F
0	0	0	0	1	1
0	0	1	0	X	1
0	1	0	X	1	0
0	1	1	0	0	0
1	0	0	1	1	X
1	0	1	1	1	X
1	1	0	0	X	X
1	1	1	X	1	X

8. For the outputs D, E, and F in Prob. 7, obtain the minimum gate realization between the SOP (NAND-NAND) and the POS (NOR-NOR) implementations. (Hint: Determine which canonical form will result in the minimum number of gates after minimization.)

9. A microprocessor has eight address lines—$A7$, $A6$, $A5$, $A4$, $A3$, $A2$, $A1$, and $A0$—and is required to address up to 16 input/output (I/O) ports. A decoder (4:16 with two active low enables), as shown, is to be used to provide the 16 active low outputs to select the input/output ports based on the least significant address bits $A3$, $A2$, $A1$, and $A0$ when the address bits $A7$, $A6$, $A5$, $A4$ = HHHH = 1111. Using a minimum number of two-input NAND gates, design the logic to accomplish this.

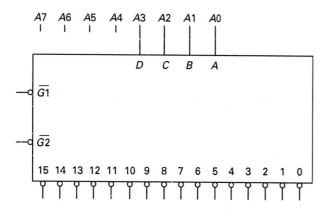

10. An input/output port selector is to use a 4:16 decoder (with two active-low enables) with the most significant bits of an address bus as shown.

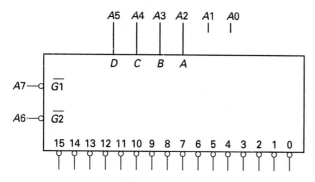

outputs Y

In addition, the selection of a port is to take place in coincidence with a timing strobe (asserted low) $\overline{X}$. Using only a quad two-input NAND gate package and a package with six NOT gates, design the circuit to select the address 00101101. Indicate which output from the 4:16 decoder ($\overline{Y}$) is to be used. Inputs to the logic are to be the correct output from the decoder, address lines $A1$ and $A0$, and the strobe $\overline{X}$.

SOLUTIONS

1. Maxterms M_1, M_2, and M_3 in binary form are given by

$$M_1 = 001$$
$$M_2 = 010$$
$$M_3 = 011$$

Adjacency exists when the terms differ by one value only. Thus, M_1 is adjacent to M_3, but not to M_2. M_2 is adjacent to M_3, but not to M_1. (This is the reason for the order of the binary numbers on the Karnaugh map, that is, 00, 01, 11, 10.)

The answer is (D).

2. Maxterms have an output value of zero. Therefore, the truth table is

F\\AB	00	01	11	10
C				
0	0			
1	0	0	0	0

The procedure for finding the minimal Karnaugh map results in the groupings shown. The maxterms M_0 and M_1 combine to give $A + B$. The bottom row combination of four variables eliminates two, specifically A and B, leaving $\overline{C}$. The resulting POS expression is

$$F(A, B, C) = \boxed{(A + B) \cdot \overline{C}}$$

The answer is (D).

3. Maxterm POS realization occurs with NOR-NOR logic as shown.

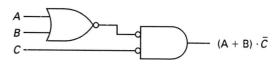

The answer is (D).

4. The minimal Karnaugh map is found using the following procedure.

step 1: Form all unique one-squares.

F\\AB	00	01	11	10
CD				
00		1		1
01	1	1	1	1
11		1	1	
10	1			

step 2: Form unique two-squares.

F AB / CD	00	01	11	10
00		[1]		[1]
01	1	[1]	1	[1]
11		1	1	
10	[1]			

step 3: Form unique four-squares.

F AB / CD	00	01	11	10
00		[1]		[1]
01	[1]	[1]	[1]	[1]
11		[1]	[1]	
10	[1]			

step 4: Form unique eight-squares.

step 5: Assign ungrouped one-squares into the largest group possible using d squares, if available. The resulting expression, taken from step 3 using the one-, two-, and four-minterm groupings is

$$F(A,B,C,D) = \boxed{\begin{array}{l} \overline{A}\,\overline{B}C\overline{D} + \overline{A}B\overline{C} \\ + A\overline{B}\,\overline{C} + \overline{C}D + BD \end{array}}$$

The answer is (C).

5.

F AB / C	00	01	11	10
0	1	1	1	1
1		d	d	1

Using the procedure for minimization of a Karnaugh map, the two groups shown are formed. Grouping the top row of minterms eliminates A and B. Grouping columns 11 and 10 eliminates B and C. The don't care condition, d, in position 111 is given the arbitrary value of one. The resulting expression is

$$F(A,B,C) = \boxed{A + \overline{C}}$$

The answer is (A).

6. The K-V map with grouping is

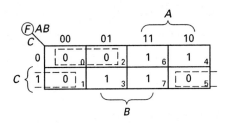

F AB / C	00	01	11	10
0	0 $_0$	0 $_2$	1 $_6$	1 $_4$
1	1 $_1$	0 $_3$	1 $_7$	0 $_5$

M_0 and M_2 combine to give $A+C$. The combination of corner maxterms M_1 and M_5 gives $B+\overline{C}$. The resulting expression is

$$F(A,B,C) = (A+C)(B+\overline{C})$$

Complementing the result gives

$$\overline{F(A,B,C)} = \overline{(A+C)(B+\overline{C})}$$

Applying De Morgan's law results in

$$\overline{F(A,B,C)} = \overline{(A+C)} + \overline{(B+\overline{C})}$$

Complementing the result puts the expression in NOR-only form.

$$F(A,B,C) = \boxed{\overline{\overline{(A+C)} + \overline{(B+\overline{C})}}}$$

The answer is (A).

7. The Karnaugh maps are as follows.

D A / BC	0	1
00	0	1
01	0	1
11	0	X
10	X	0

E A / BC	0	1
00	1	1
01	X	1
11	0	1
10	1	X

F A / BC	0	1
00	1	X
01	1	X
11	0	X
10	0	X

The Veitch diagrams are as follows.

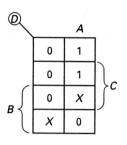
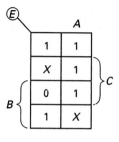

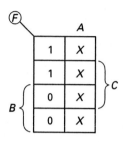

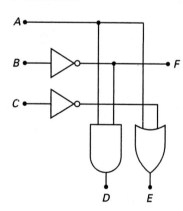

8. After minimization, the output is

$$D = A \cdot \overline{B}$$

$$\overline{E} = \overline{A} \cdot C$$

$$E = A + \overline{C}$$

$$F = \overline{B}$$

The circuit is realized as

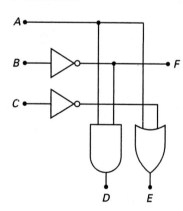

9. The outputs of the decoder are all high, except for the case when $\overline{G}1$ and $\overline{G}2$ are both low. In that case, the output with the binary address corresponding to $A3$, $A2$, $A1$, and $A0$ will go low, thus selecting a particular device.

The objective is to provide logic, using a minimum number of two-input NAND gates, so that the decoder will be active when lines $A4$, $A5$, $A6$, and $A7$ are all high.

This can be done with two NAND gates as shown.

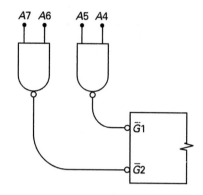

10. The address used is

$$A_7 A_6 A_5 A_4 A_3 A_2 A_1 A_0 = 00101101$$

$$A_5 A_4 A_3 A_2 = 1011_2 = 11_{10}$$

The output selected is Y_{11}. Y_{11} is asserted low. The desired logic function is

$$F = \overline{X}\,\overline{Y}_{11}\overline{A}_1 A_0$$

The realization with two input gates follows.

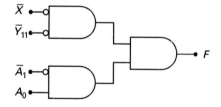

The NAND/NOT realization follows.

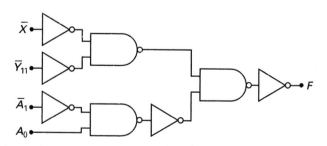

52 Synchronous Sequential Networks

PRACTICE PROBLEMS

1. A register is
(A) a flip-flop
(B) a set of flip-flops
(C) a memory device
(D) both (B) and (C)

2. What combination of values is NOT allowed as input to an S-R flip-flop?
(A) $S = 0, R = 0$
(B) $S = 0, R = 1$
(C) $S = 1, R = 0$
(D) $S = 1, R = 1$

3. In the S-R flip-flop timing diagram shown, what is the output sequence from point 1 to point 3?

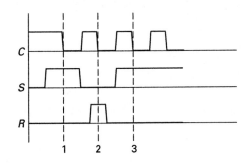

(A) 000
(B) 010
(C) 101
(D) 111

4. What best describes the overall function of the network shown?

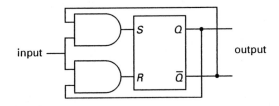

(A) delay
(B) set-reset/single input
(C) toggle
(D) none of the above

5. Of the four principal flip-flops, which can be used as the building block for all the others?
(A) D
(B) J-K
(C) S-R
(D) T

6. Develop the state transition diagram for the synchronous sequential circuit shown. Show all work.

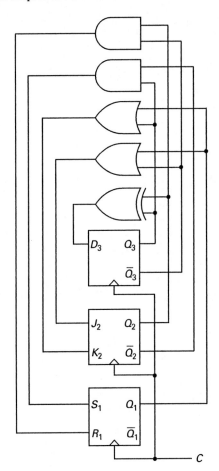

7. For the counter circuit shown, develop the state table and the state transition diagram.

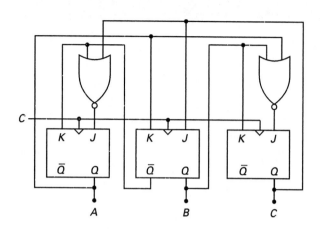

8. Using T flip-flops, design a counter that will produce the following sequence.

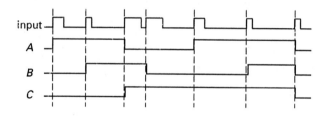

9. Four-bit binary ripple counters are available. Each has parallel outputs $DCBA$, with A the least significant bit (LSB), a count input (CI), and a carry out (CO) that is low except when the counter output is $DCBA = HHHH$. The ripple counters also have a level-sensitive preset input (PR), which will set $DCBA = HHHH$ when PR is high. The counters respond to the falling edge of the count input.

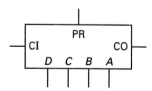

Using two such counters and necessary logic gates, design a circuit that will provide a one (high) after each 100 counts.

10. For the digital circuit shown, complete the state diagram. Show all transitions and the method of analysis.

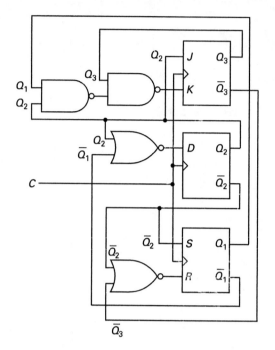

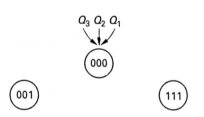

11. A counter is constructed from J-K master-slave flip-flops and NAND gates. Show one complete cycle of the counter. Assume $Q_1 = Q_2 = Q_3 = 0$ initially.

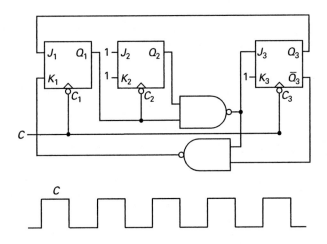

12. Design a digital circuit using only two-input NAND gates and J-K master-slave flip-flops. Use as many of each as is necessary, but the design should be minimized with respect to packages. (There are four two-input NAND gates in a package and two J-K flip-flops per package.)

The system is to have two inputs: a clock and X. The X is a sequence of binary voltages—high = 1 and low = 0. Binary data is presented in sequences of one, two, or three ones followed by at least one zero. No other sequence will ever occur (i.e., a sequence of four ones will not occur).

There are to be two outputs, A and B. A is to go high when the first one of a sequence occurs and go low after the second or when the next zero occurs. B is to go high when the second one of a sequence occurs and stay high until the next zero occurs.

13. A shaft is encoded by having a conductor covering from 0° to 180° and an insulator covering from 180° to 360°. A 5 V source is connected to the conductor. One stationary carbon brush sensor contacts the shaft at a reference position, and another contacts the shaft 90° from the first in the positive direction of shaft rotation. When a brush is in contact with the metal sector of the shaft, it delivers a high (approximately 5 V) voltage to one input of the logic circuit, and when it is not in contact with the metal sector, it provides 0 V to that input.

A sequential logic circuit is to be clocked at a frequency more than ten times the maximum shaft rotational velocity and is to provide input signals to an up-down counter to correspond with the number of rotations the shaft makes. The counter output is then to be the integral number of rotations the shaft makes.

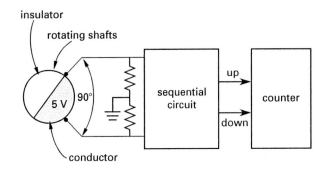

14. The following flowchart deals with three input variables (A, B, and C) and three output variables (X, Y, and Z). Using two-input NAND gates, design combinational logic to realize the flowchart functions.

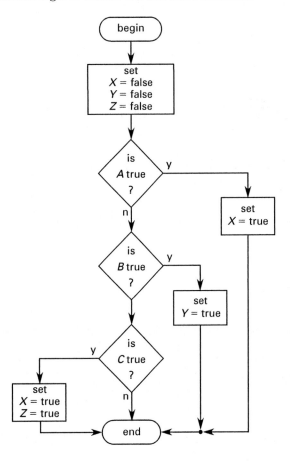

SOLUTIONS

1. A register is the computer hardware required to store one machine word. It is used as memory in synchronous sequential networks and is constructed from a set of flip-flops.

The answer is (D).

2. The $S = 1$ and $R = 1$ state in an S-R flip-flop is disallowed because the output is indeterminate.

The answer is (D).

3.

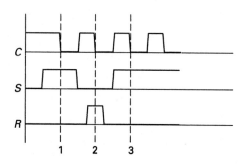

At point 1, $S = 1$, $R = 0$, the flip-flop is set, and $Q = 1$. At point 2, $S = 0$, $R = 1$, the flip-flop is reset, and $Q = 0$. At point 3, $S = 1$, $R = 0$, the flip-flop is set, and $Q = 1$. The output sequence is 101.

The answer is (C).

4. The combination of an S-R flip-flop and logic elements connected as indicated forms a toggle. (This can be discerned by assuming an output, say $Q = 0$, and then varying the input to determine the response. The process is repeated for $Q = 1$. The truth table for the network can then be constructed and the overall response determined.)

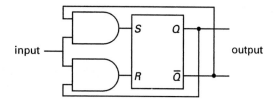

The answer is (C).

5. Of the four principal flip-flops, all can be constructed using an S-R flip-flop as the building block. Thus, an S-R flip-flop requires no additional logic elements to construct.

The answer is (C).

6. Tracing the circuit results in the following equations.

$$R_1 = Q_2 \cdot \overline{Q_3}$$
$$S_1 = \overline{Q_2} \cdot Q_3$$
$$K_2 = Q_1 + Q_3$$
$$J_2 = Q_1 + \overline{Q_3}$$
$$D_3 = Q_2 \oplus Q_3$$

The exitation table is

Q_1	Q_2	Q_3	R_1	S_1	K_2	J_2	D_3	Q_1^+	Q_2^+	Q_3^+
0	0	0	0	0	0	1	0	0	1	0
0	0	1	0	1	1	0	1	1	0	1
0	1	0	1	0	0	1	1	0	1	1
0	1	1	0	0	1	0	0	0	0	0
1	0	0	0	0	1	1	0	1	1	0
1	0	1	0	1	1	1	1	1	1	1
1	1	0	1	0	1	1	1	0	0	1
1	1	1	0	0	1	1	0	1	0	0

The state transition diagram can be taken directly from the table as

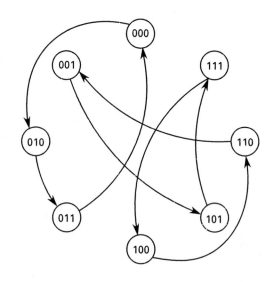

7. Tracing the circuit results in the following equations.

$$K_A = \overline{B}$$

$$K_B = A$$

$$K_C = B$$

$$J_A = \overline{\overline{B} + C} = B \cdot \overline{C}$$

$$J_B = C$$

$$J_C = \overline{A + B} = \overline{A} \cdot \overline{B}$$

The state table for a J-K flip-flop is

J	K	Q	Q^+
0	0	0	0
0	0	1	1
0	1	0	0
0	1	1	0
1	0	0	1
1	0	1	1
1	1	0	1
1	1	1	0

The next state for a J-K is given by

$$Q^+ = Q \cdot \overline{K} + \overline{Q} \cdot J$$

Thus, the outputs A, B, and C are

$$A^+ = A \cdot B + \overline{A}B\overline{C} = A \cdot B + B \cdot \overline{C}$$

$$= (110) \text{ or } (111) \text{ or } (010)$$

$$B^+ = \overline{A} \cdot B + \overline{B}C$$

$$= (010) \text{ or } (011) \text{ or } (001) \text{ or } (101)$$

$$C^+ = \overline{B}C + \overline{A} \cdot \overline{B} \cdot \overline{C} = \overline{B}C + \overline{A} \cdot \overline{B}$$

$$= (000) \text{ or } (001) \text{ or } (101)$$

The state table for the counter is

A	B	C	A^+	B^+	C^+
0	0	0	0	0	1
0	0	1	0	1	1
0	1	0	1	1	0
0	1	1	0	1	0
1	0	0	0	0	0
1	0	1	0	1	1
1	1	0	1	0	0
1	1	1	1	0	0

The state transition diagram follows.

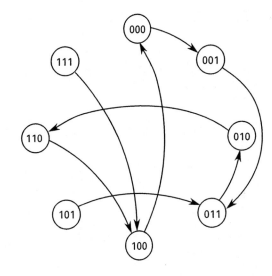

8. From the timing diagram, the sequence triggered by the rising clock edge is

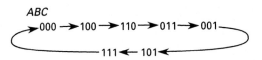

The next-state table is

A	B	C	A^+	B^+	C^+	T_A	T_B	T_C
0	0	0	1	0	0	1	0	0
0	0	1	1	0	1	1	0	0
0	1	0	–	–	–	–	–	–
0	1	1	0	0	1	0	1	0
1	0	0	1	1	0	0	1	0
1	0	1	1	1	1	0	1	0
1	1	0	0	1	1	1	0	1
1	1	1	0	0	0	1	1	1

From this,

$$T_A = \overline{A} \cdot \overline{B} + A \cdot B$$

$$T_B = A \cdot \overline{B} + BC$$

$$T_C = A \cdot B$$

This requires six two-input NAND gates.

Something must be done about the 010 state, because it results in a hang-up state if it is the state in existence at start-up. This can be done by adding a term $B \cdot \overline{C}$ to T_C.

$$T_C = A \cdot B + B \cdot \overline{C}$$

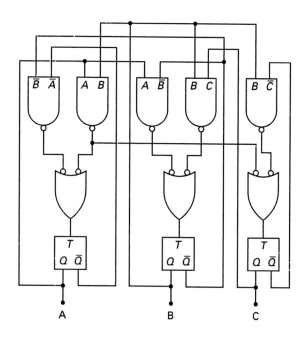

9. Preset must be activated at 99, not 100.

$$99_{10} = 64 + 32 + 2 + 1 = 01100011_2$$

Use an AND gate to obtain the needed preset.

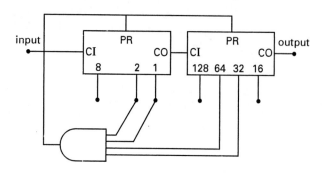

10. The logic for the flip-flop inputs is

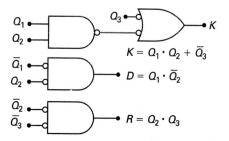

$$K = Q_1 \cdot Q_2 + \overline{Q}_3$$
$$D = Q_1 \cdot \overline{Q}_2$$
$$R = Q_2 \cdot Q_3$$

The equations for all the inputs are

$$J = Q_2$$

$$K = Q_1 \cdot Q_2 + \overline{Q}_3$$

$$D = Q_1 \cdot \overline{Q}_2$$

$$S = \overline{Q}_2$$

$$R = Q_2 \cdot Q_3$$

The excitation table is

Q_3	Q_2	Q_1	K	J	D	R	S	Q_3^+	Q_2^+	Q_1^+
0	0	0	1	0	0	0	1	0	0	1
0	0	1	1	0	1	0	1	0	1	1
0	1	0	1	1	0	0	0	1	0	0
0	1	1	1	1	0	0	0	1	0	1
1	0	0	0	0	0	0	1	1	0	1
1	0	1	0	0	1	0	1	1	1	1
1	1	0	0	1	0	1	0	1	0	0
1	1	1	1	1	0	1	0	0	0	0

The state diagram taken directly from the table must be

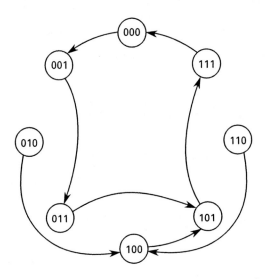

The state diagram is taken from the table.

$Q_1 \ Q_2 \ Q_3$

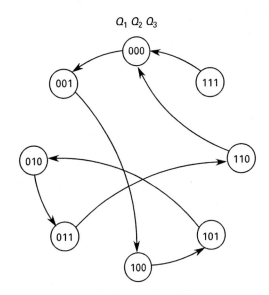

11. Q_2 will toggle when Q_1 falls from high to low. (Outputs change on the falling edge of the clock.)

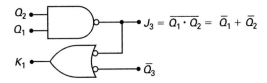

$J_3 = \overline{Q_1 \cdot Q_2} = \overline{Q}_1 + \overline{Q}_2$

$\overline{Q}_3$

The flip-flop input equations are

$$J_1 = Q_3$$

$$K_1 = Q_1 \cdot Q_2 + Q_3$$

$$J_3 = \overline{Q}_1 + \overline{Q}_2$$

$$K_3 = 1$$

The timing diagram that correlates to the state diagram follows.

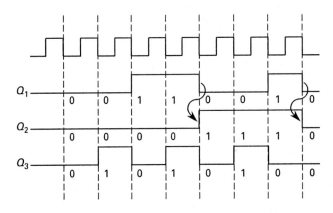

The excitation table is

$Q_1 \ Q_2 \ Q_3$	$J_1 \ K_1$	$Q_1 \to Q_1^+$	J_3	$Q_1^+ \ Q_2^+ \ Q_3^+$
0 0 0	0 0	0→0	1	0 0 1
0 0 1	1 1	0→1	1	1 0 0
0 1 0	0 0	0→0	1	0 1 1
0 1 1	1 1	0→1	1	1 1 0
1 0 0	0 0	1→1	1	1 0 1
1 0 1	1 1	1→0	1	0 1 0
1 1 0	0 1	1→0	0	0 0 0
1 1 1	1 1	1→0	0	0 0 0

12. The state diagram is

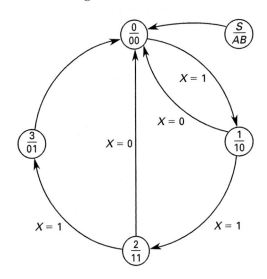

There are four states, so two flip-flops are required. The states are set to minimize output logic.

	A	B	X	A^+	B^+	J_A	K_A	J_B	K_B
S_0	0	0	0	0	0	0	X	0	X
			1	1	0	1	X	0	X
S_3	0	1	0	0	0	0	X	X	1
			1	0	0	0	X	X	1
S_1	1	0	0	0	0	X	1	0	X
			1	1	1	X	0	1	X
S_2	1	1	0	0	0	X	1	X	1
			1	0	1	X	1	X	0

The flip-flop inputs are determined from Veitch diagrams.

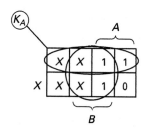

$$J_A = \bar{B} \cdot X$$

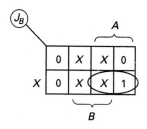

$$K_A = B + \bar{X}$$

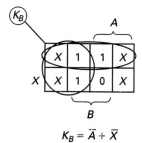

$$J_B = A \cdot X$$

$$K_B = \bar{A} + \bar{X}$$

The resulting circuit follows.

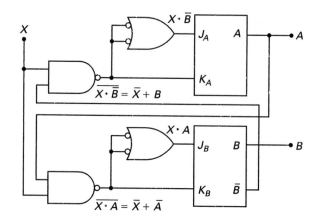

13. The rotating shaft provides the following inputs.

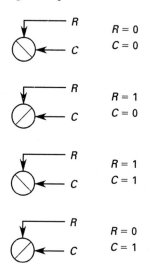

For a positive sequence the resulting input is

R	0 0 0 1 1 1 1 0 0 0 0
C	0 0 0 0 0 1 1 1 1 0 0

For a negative sequence the resulting input is

R	0 0 0 0 1 1 1 1 1 1 0 0 0 0
C	0 1 1 1 1 1 1 0 0 0 0 0 0 0

One solution assigns states to correspond to the four possible combinations of R and C.

S_0 (R, C) has been (00)

S_1 (R, C) has been (01)

S_2 (R, C) has been (10)

S_3 (R, C) has been (11)

The outputs are assigned to the transition between two states, say S_1 and S_3. When the transition is from S_1 to S_3, the output is up. When the transition is from S_3 to S_1, the output is down.

The transition from S_1 to S_3 occurs when

Q_R	Q_C	R	C
0	1	1	1

$$\text{up} = \overline{Q}_R Q_C R C$$

The transition from S_3 to S_1 occurs when

Q_R	Q_C	R	C
1	1	0	1

$$\text{down} = Q_R Q_C \overline{R} C$$

The transition table is

S	Q_R	Q_C	R	C	Q_R^+	Q_C^+	up	down
S_0	0	0	0	0	0	0	0	0
S_0	0	0	0	1	0	1	0	0
S_0	0	0	1	0	1	0	0	0
S_0	0	0	1	1	X	X	0	0
S_1	0	1	0	0	0	0	0	0
S_1	0	1	0	1	0	1	0	0
S_1	0	1	1	0	X	X	0	0
S_1	0	1	1	1	1	1	1	0
S_2	1	0	0	0	0	0	0	0
S_2	1	0	0	1	X	X	0	0
S_2	1	0	1	0	1	0	0	0
S_2	1	0	1	1	1	1	0	0
S_3	1	1	0	0	X	X	0	0
S_3	1	1	0	1	0	1	0	1
S_3	1	1	1	0	1	0	0	0
S_3	1	1	1	1	1	1	0	0

From this, $D_R = R$, and $D_C = C$.

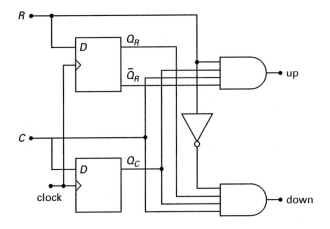

14. From the flow chart, the resulting equations are derived.

$$X = A + \overline{B} \cdot C$$

$$Y = \overline{A} \cdot B$$

$$Z = \overline{A} \cdot \overline{B} \cdot C$$

The logic to realize the equations follows.

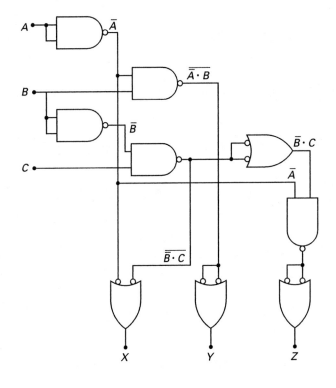

53 Programming Languages

PRACTICE PROBLEMS

1. If J = 3, K = 4, L = 6, and M = 9, what is I = (J*K) + (L − M)?

2. If J = 15, K = 4, L = 9, and M = 19, what is the value of the logical variable X?

(a) X = J.LT.L.OR.K.GE.(M−J)

(b) X = J.GT.L.OR.K.LT.(M−J)

3. What is the value of Q?

```
            R = 18.
            S = 6.
            T = 2.
            Q = R/S**T−T
            IF Q 10, 20, 30
      10    Q = 10
      15    GO TO 40
      20    Q = 100
      25    GO TO 40
      30    Q = 1000
      40    END
```

4. Write a DO loop to multiply ten numbers together until the product exceeds five.

5. Write a DO loop to add the remainders when each of ten numbers is divided by seven.

6. Write the FORTRAN statements that will print the numbers 4104, 6802, and 5798 in scientific notation with two digits to the right of the decimal point, leaving one line between each number.

7. What is the output of the following routine?

```
            F = 50.0*4.6
            WRITE (6,100)F
      100   FORMAT (('  '),T12,E7.3)
```

8. Write a statement function to find the sine of θ, given that $\sin\theta = \sqrt{\omega t}$.

9. Write a statement function to convert inches to feet.

10. Write a subroutine that can look at a list of numbers and sum only those that are greater than or equal to ten.

11. A main program consists of the following statements.

```
            COMMON ALPHA, BETA
            ALPHA = 4.0
            BETA = 3.0
```

What must be written in the beginning of the subroutine so that A takes the value of ALPHA in the main program and B takes the value of BETA in the main program?

12. A, B, and C are positive, unequal integers. Write a FORTRAN program to find and print out the smallest integer.

13. A program is to calculate the force required to move springs various distances, using the formula $F = kx$. For each calculation, the user will enter the spring constant and distance. Write a program that performs the calculation, and then prints the results to look like the following sample.

K = 01.300   X = 14.500   F = 18.900

Some lines have been written already.

```
            DATA K,X/1.3,14.5
            READ (5,100)K,X

            F = K*X

      999   END
```

14. Consider the C++ programming statement shown.

```
# include <iostream>
using namespace std;
```

PROFESSIONAL PUBLICATIONS, INC.

Computers

What is the purpose of the programming statement?
- (A) allows input to iostream
- (B) allows input to namespace
- (C) makes iostream file available
- (D) opens input/output interface

15. Which of the following represents the extraction (or input) operator in C++?
- (A) `<<`
- (B) `>>`
- (C) `<=`
- (D) `=>`

16. Let a = 20 and b = 15. What is the value of the variable a following the C++ programming operation shown?

$$a \mathrel{-}= b$$

- (A) −15
- (B) −5
- (C) 5
- (D) 35

17. Let a = 10. What value is stored in a following the C++ operation ++a?
- (A) 8
- (B) 9
- (C) 10
- (D) 11

18. Which of the following is a literal in the C++ programming language?
- (A) `()`
- (B) `{}`
- (C) `and`
- (D) `5`

19. What does the ∥ operator indicate in the C++ programming language?
- (A) AND
- (B) OR
- (C) parallel
- (D) true if

20. The logic of a C++ program resides in the control structures. Which of the following is NOT a control structure?
- (A) `for` loop
- (B) `if/else` statement
- (C) `long` modifier
- (D) `while` statement

SOLUTIONS

1. The algebraic translation is

$$I = JK + L - M = (3)(4) + 6 - 9$$
$$= \boxed{9}$$

2. The algebraic translations are

(a) X is true if J < L or K ≥ (M − J). Otherwise, X is false.

Because K is equal to (M − J),

$$\boxed{\text{X is true.}}$$

(b) X is true if J > L or K < (M − J). Otherwise, X is false.

Because J is greater than L,

$$\boxed{\text{X is true.}}$$

3. The algebraic translation of the FORTRAN formula for Q is

$$Q = \frac{R}{S^T} - T = \frac{18}{(6)^2} - 2 = -1.5$$

Because Q is negative, go to statement 10.

$$10 \quad Q = 10$$

Go to statement 40.

$$40 \quad \text{END}$$

Because the program ends at statement 40 with Q still equal to 10,

$$\boxed{Q = 10.}$$

4.
```
    PROD=1.0
    DO 15 I=1, 10
    READ(5,10) B
10  FORMAT(F7.4)
    PROD=PROD*B
    IF (PROD .GT. 5.0) GO TO 20
15  CONTINUE
20  (any subsequent statement)
```

5.
```
        SUM=0.0
        DO 15 I=1,10
        READ(5,10) B
     10 FORMAT(F7.4)
        C=AMOD(B/7.0)
        SUM=SUM+C
     15 CONTINUE
```

6.
```
        A = 4104.
        B = 6802.
        C = 5798.
        WRITE(6,10) A,B,C
     10 FORMAT('',3(E7.2/))
```

7. 2.300 E2 (starting in column 12)

8.
```
        REAL FUNCTION STHETA (OMEGA,T)
        STHETA = SQRT (OMEGA*T)
        RETURN
        END
```

9.
```
        REAL FUNCTION FEET(RINCHS)
        FEET = RINCHS/12.0
        RETURN
        END
```

10.
```
        SUBROUTINE ADD (B , SUM)
        IF (B.LT.10.0) GO TO 5
        SUM = SUM+B
      5 RETURN
        END
```

11.
```
        COMMON A,B
```

12.
```
        INTEGER A,B,C
        READ (5,10) A,B,C
     10 FORMAT (3I4)
        IF(A.LT.B.AND.A.LT.C) I=A
        IF(B.LT.C.AND.B.LT.A) I=B
        IF(C.LT.A.AND.C.LT.B) I=C
        WRITE(6,20) I
     20 FORMAT('','THE SMALLEST
              INTEGER IS',I4)
        STOP
        END
```

13.
```
        REAL K
        DATA K,X/1.3,14.5
        F=K*X+0.05
        F=0.1*FLOAT(INT(F*10.0))
        WRITE(6,200) K,X,F
    200 FORMAT('','K=',F6.3,'','X=',F6.3,
             '','F=',F6.3)
        STOP
    999 END
```

An equivalent program can be written by substituting the following code for the data statement.

```
        READ(5,100) K,X
    100 FORMAT(2F4.1)
```

14. The `# include` is called an "include" statement. This causes the program to access the `iostream` file from the C++ standard library.

The answer is (C).

15. The input operator in C++ is `>>`.

The answer is (B).

16. The operator `-=` results in subtraction and assignment. That is,

$$a = a - b$$
$$= 20 - 15$$
$$= \boxed{5}$$

The answer is (C).

17. The `++a` operation in C++ increments the value in a and then returns the incremented value. Thus,

```
a   = 10
++a = 11
a   = 11
```

The answer is (D).

18. A literal is a symbol for a specific value of a variable. The "( )" and "{ }" are predefined symbols. The term `and` is a keyword. The number `5` is the only possible literal.

The answer is (D).

19. The double pipe, ‖, is the OR operator in C++. Thus, a‖b returns `true` if a OR b is `true`.

The answer is (B).

20. The `long` modifier is a data type. All the other items are control structures that return `true` or `false` upon execution.

The answer is (C).

54 Analog Computers: Analog Logic Functions

PRACTICE PROBLEMS

1. Draw the analog computer circuit for the given differential equation.

$$V_0 = -\int (V_1 + 4V_2)\, dt$$

2. Draw the analog simulation for the given equation.

$$7x''' + 14x'' + 42x' + 2x = 49$$

3. What is the analog computer simulation for the mechanical system shown?

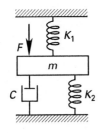

4. Draw an analog circuit to simulate variable x in the mechanical system shown.

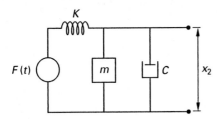

5. Draw an analog circuit to simulate variable x in the mechanical system shown.

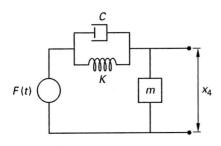

SOLUTIONS

1. Differentiate both sides.

$$V_0' = -V_1 - 4V_2$$
$$-V_0' = V_1 + 4V_2$$

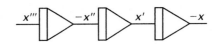

2. *step 1:* $\qquad x''' + 2x'' + 6x' + \frac{2}{7}x = 7$

step 2: $\quad -x''' = 2x'' + 6x' + \frac{2}{7}x - 7$

step 3:

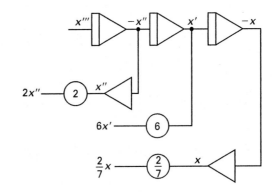

step 4:

steps 5 and 6:

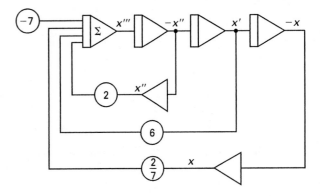

3.

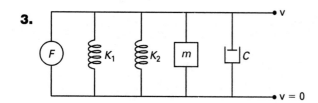

4.

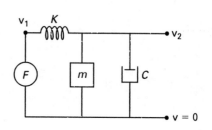

$$F = ma + Cv + (K_1 + K_2)x$$

$$mx'' + Cx' + (K_1 + K_2)x - F = 0$$

$$F = ma_2 + Cv_2$$

$$mx_2'' + Cx_2' - F = 0$$

step 1: $\quad x'' + \left(\dfrac{C}{m}\right)x' + \left(\dfrac{K_1 + K_2}{m}\right)x - \dfrac{F}{m} = 0$

step 2: $\quad -x'' = \left(\dfrac{C}{m}\right)x' + \left(\dfrac{K_1 + K_2}{m}\right)x - \dfrac{F}{m}$

step 3:

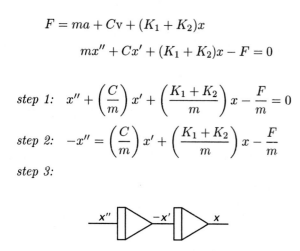

step 1: $\quad x_2'' + \left(\dfrac{C}{m}\right)x_2' - \dfrac{F}{m} = 0$

step 2: $\quad -x_2'' = \left(\dfrac{C}{m}\right)x_2' - \dfrac{F}{m}$

step 3:

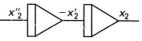

step 4:

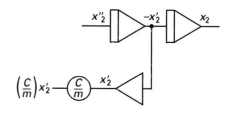

step 4:

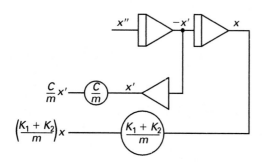

steps 5 and 6:

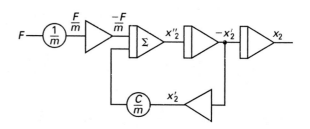

Notice that $F = K(x_1 - x_2)$ is a second equation that applies to this system. Because x_2 is available, x_1 can be found.

$$x_1 = \frac{F}{K} + x_2$$

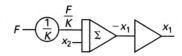

steps 5 and 6:

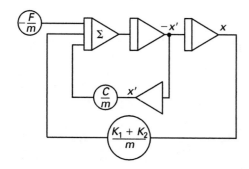

The entire system is

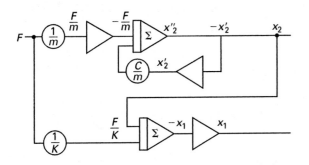

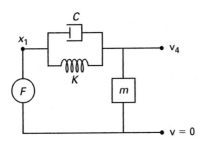

5.

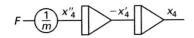

$$F = ma_4$$

$$mx_4'' - F = 0$$

step 1: $\quad x_4'' - \dfrac{F}{m} = 0$

step 2: $\quad x_4'' = \dfrac{F}{m}$

step 3:

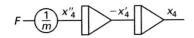

Notice that $F = C(v_1 - v_4) + K(x_1 - x_4)$ is a second equation that applies to the system. Because x_4 and x_4' are available, x_1 can be found.

$$F = Cx_1' - Cx_4' + Kx_1 - Kx_4$$

$$-x_1' = -x_4' + \left(\dfrac{K}{C}\right)x_1 - \left(\dfrac{K}{C}\right)x_4 - \dfrac{F}{C}$$

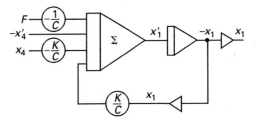

The entire system is

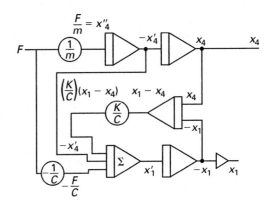

55 Operating Systems, Networking Systems, and Standards

PRACTICE PROBLEMS

1. Which portion of the operating system allocates resources, responds to service calls, and is considered the core of the operating system?
- (A) driver
- (B) kernel
- (C) library
- (D) memory

2. Which of the following is a term for a performance test used to compare software programs?
- (A) benchmark
- (B) engineering requirement
- (C) request for comment (RFC)
- (D) standard

3. Consider the illustration shown. What type of topology is represented?

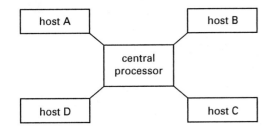

- (A) bus
- (B) ring
- (C) star
- (D) tree

4. When a dedicated path exists between computer networks, the communication is referred to as _____ switching.
- (A) circuit
- (B) computer
- (C) dedicated
- (D) packet

5. The set of hardware and software interfaces, or rules, that govern the transfer of data within a network is referred to as
- (A) domains
- (B) protocols
- (C) routers
- (D) standards

SOLUTIONS

1. The core of any operating system is the kernel.

The answer is (B).

2. A benchmark is a test of computer hardware or software that is generally run on a number of products to compare performance.

The answer is (A).

3. Network topologies include bus, star, ring, and tree. The illustration shown is the star topology.

The answer is (C).

4. A dedicated path of fixed bandwidth and speed in network communication is an example of circuit switching.

The answer is (A).

5. The transfer of data is governed by protocols—some of which can be standards, but are not necessarily so.

The answer is (B).

56 Digital Systems

PRACTICE PROBLEMS

1. Which of the following components does NOT exist in a digital system?
- (A) central processing unit
- (B) control unit
- (C) information processing unit
- (D) register

2. What term is used for a synchronizing element that temporarily stores information during interactions between electronic devices?
- (A) buffer
- (B) flag
- (C) handshaking unit
- (D) interface

3. What is the maximum value that can be represented by a 4-bit binary code?
- (A) 8
- (B) 15
- (C) 16
- (D) 32

4. What is the voltage per step of an A/D converter with an input voltage range of −5 V to +5 V and a 4-bit output?
- (A) 0.32 V/step
- (B) 0.34 V/step
- (C) 0.63 V/step
- (D) 0.67 V/step

5. What is the magnitude of the maximum output voltage of an inverting ladder-type D/A converter operating as a 4-bit system and using a 10 V reference voltage?
- (A) 9.40 V
- (B) 16.0 V
- (C) 18.8 V
- (D) 19.4 V

6. A digitally controlled switch is shown. When the binary input is logic low, what is the state of the PMOS?

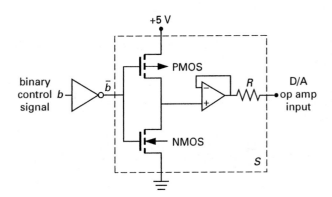

- (A) indeterminate
- (B) off
- (C) on
- (D) same as the NMOS

7. The A/D converter shown uses a 3 MHz clock and a 5 V reference voltage. Determine the maximum number of bits for the binary counter if the conversion is to be completed in less than 1 ms. (a) Determine the value of RC to permit full range of the integrator output to −10 V. (b) What is the fundamental accuracy of the conversion of an analog voltage?

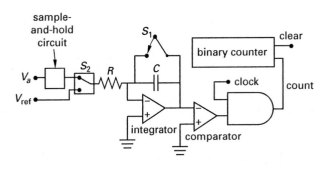

8. The circuit shown is a programmable gain amplifier (PGA). It is to be programmable up to a maximum gain of 2.00. It is to be controlled by an 8-bit binary coded data word that controls the electronic switches S_0 through S_7, with S_7 being the most significant bit. Specify the resistance values needed to accomplish this.

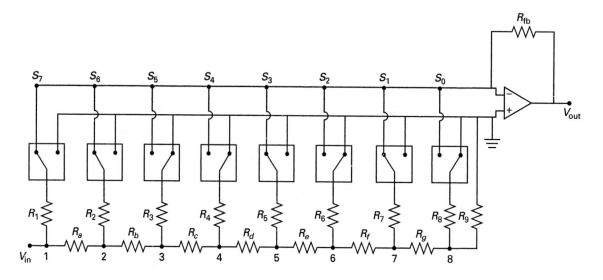

SOLUTIONS

1. The central processing unit, or CPU, is used in computers to allow the system to respond to variations in programming. The digital system has no such unit.

The answer is (A).

2. Temporary storage of information during interactions between electronic devices occurs in memory elements called buffers.

The answer is (A).

3. The maximum value is given by

$$\text{maximum value} = 2\text{MSB} - \text{LSB}$$

$$= (2)(2)^3 - (2)^0$$

$$= \boxed{15}$$

The answer is (B).

4. The voltage range is 10 V. The number of steps for an n-bit number is given by $2^n - 1$. Thus,

$$\frac{\text{volts}}{\text{step}} = \frac{10\text{ V}}{(2)^n - 1} = \frac{10\text{ V}}{(2)^4 - 1} = \frac{10\text{ V}}{15}$$

$$= \boxed{0.67\text{ V/step}}$$

The answer is (D).

5. The maximum voltage for an n-bit D/A converter is given by

$$|V_{\text{out,max}}| = V_{\text{ref}} \left(2 - \frac{1}{(2)^{n-1}} \right) = (10\text{ V}) \left(2 - \frac{1}{(2)^{4-1}} \right)$$

$$= (10\text{ V}) \left(2 - \frac{1}{8} \right)$$

$$= \boxed{18.75\text{ V} \quad (18.8\text{ V})}$$

The answer is (C).

6.

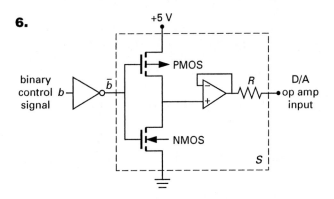

When b is logic low, $\bar{b}$ is logic high. The PMOS is an enhancement type. The $\bar{b}$ signal is a positive voltage on the gate. A negative voltage is required to create the conductivity channel for hole carriers. Therefore, the PMOS is off. (The NMOS is on, placing 0 V on the noninverting terminal and causing the op amp output to be low.)

The answer is (B).

7. (a) In 1 ms, a 3 MHz clock goes through 3000 clock pulses. The closest binary number less than 3000 is 2048, so the full count will be 2048. This is obtained with an 11-bit counter with place values.

$$(2)^{10} + (2)^9 + (2)^8 + (2)^7 + (2)^6 + (2)^5$$
$$+ (2)^4 + (2)^3 + (2)^2 + (2)^1 + 1 = (2)^{11} - 1$$

The duration of a full count is

$$\frac{(2)^{11}\text{ counts}}{3 \times 10^6\ \dfrac{\text{counts}}{\text{s}}} = 682.7 \times 10^{-6}\text{ s}$$

$$\frac{v_{\text{full range}}}{RC} \int_0^{682.7 \times 10^{-6}\text{ s}} dt = v_{\text{max}}$$

$$RC = \left(\frac{v_{\text{full range}}}{v_{\text{max}}} \right)$$

$$\times (682.7 \times 10^{-6}\text{ s})$$

For $|v_{\text{full range}}| = |v_{\text{max}}| = 10$ V,

$$\boxed{RC = 682.7 \times 10^{-6}\text{ s}}$$

(b) The accuracy is $\pm\ ^1\!/_2$ least significant bit (LSB), and the LSB has a value of $10\text{ V}/(2)^{11}$, so the accuracy is

$$\pm \tfrac{1}{2}(\text{LSB}) = \left(\frac{1}{2} \right) \left(\frac{10\text{ V}}{(2)^{11}} \right) = \frac{10\text{ V}}{(2)^{12}}$$

$$= \boxed{2.44\text{ mV}}$$

Computers

8. The resistors R_1 through R_8 are either connected to ground directly at the positive terminal of the op amp or connected to ground effectively via the negative terminal. In either case, the impedance of the input circuit remains unchanged by the switch positions.

The overall gain, with all switches to the left, is to be two, and the R_1 contribution is to be one.

$$-v_{\text{out}} = v_{\text{in}}\left(\frac{R_{\text{fb}}}{R_1}\right) + v_2\left(\frac{R_{\text{fb}}}{R_2}\right) + v_3\left(\frac{R_{\text{fb}}}{R_3}\right) + \cdots$$

$$\cdots + v_6\left(\frac{R_{\text{fb}}}{R_6}\right) + v_7\left(\frac{R_{\text{fb}}}{R_7}\right) + v_8\left(\frac{R_{\text{fb}}}{R_8}\right)$$

The most straightforward solution is to make $R_1 = R_2 = \cdots = R_7 = R_{\text{fb}}$, and then to design the circuit so that

$$v_2 = \frac{v_{\text{in}}}{2}$$

$$v_3 = \frac{v_{\text{in}}}{4}$$

$$v_4 = \frac{v_{\text{in}}}{8}$$

$$\vdots$$

$$v_8 = \frac{v_{\text{in}}}{128}$$

At the LSB position,

$$v_8 = \frac{v_{\text{in}}}{128}$$

$$i_g = \left(\frac{v_{\text{in}}}{128}\right)\left(\frac{1}{R_8} + \frac{1}{R_9}\right)$$

If $R_8 = R_9 = R$,

$$i_g = \frac{v_{\text{in}}}{64R}$$

$$v_7 = v_8 + R_g i_g = \frac{v_{\text{in}}}{128} + \frac{2R_g v_{\text{in}}}{128R}$$

$$= \left(\frac{v_{\text{in}}}{128}\right)\left(1 + \frac{2R_g}{R}\right) \to \frac{v_{\text{in}}}{64}$$

$$R_g = \frac{R}{2}$$

Then,

$$i_f = i_g + \frac{v_7}{R}$$

$$= \frac{v_{\text{in}}}{64R} + \left(\frac{1}{R}\right)\left(\frac{v_{\text{in}}}{64}\right) = \frac{v_{\text{in}}}{32R}$$

$$v_6 = v_7 + R_f i_f = \frac{v_{\text{in}}}{64} + \left(\frac{R_f}{32R}\right)v_{\text{in}}$$

$$= \left(\frac{v_{\text{in}}}{64}\right)\left(1 + \frac{2R_f}{R}\right) \to \frac{v_{\text{in}}}{32}$$

Then, $R_f = R/2 = R_g$.

This process can be repeated to show

$$R_g, R_f, \ldots, R_a = \frac{R}{2}$$

$$R_a = R_b = R_c = R_d = R_e = R_f = \boxed{R_{\text{fb}}/2}$$

$$R_1 = R_2 = R_3 = R_4 = R_5 = R_6 = R_7 = R_8 = \boxed{R_{\text{fb}}}$$

High-Frequency Transmission

PRACTICE PROBLEMS

1. A transmission line with a characteristic impedance of 50 Ω and a termination resistance of 70 Ω has $\beta l = \pi/2$. Determine the (a) reflection coefficient, (b) input impedance, and (c) VSWR.

2. A transmission line with $Z_0 = 50$ Ω is terminated with a load impedance of $25 - j25$ Ω. (a) Find the electrical angle (βl) where a compensating capacitor can be inserted in the line to cause the impedance seen at that point to be 50 Ω. (b) Determine the reactance of the capacitor.

3. Given a coaxial cable with $Z_0 = 50$ Ω, determine the length of cable (in wavelengths) necessary to produce (a) a reactance of -25 Ω, and (b) a reactance of $+50$ Ω. (The stubs may be either open or shorted.)

4. A transmission line with $Z_0 = 72$ Ω and $\beta = 0.5$ rad/m is 7.5 m long. It is terminated with a pure 50 Ω resistance. Determine the impedance seen from the source.

5. A transmission line with $Z_0 = 100$ Ω is terminated with $Z_L = 25 + j25$ Ω. This is to be compensated with a 72 Ω line section. Determine the length of the compensating line section, and specify its placement from the load on the 100 Ω line for (a) series compensation and (b) parallel compensation.

SOLUTIONS

1. (a) $\Gamma_{\text{load}} = \dfrac{Z_L - Z_0}{Z_L + Z_0} = \dfrac{70 \ \Omega - 50 \ \Omega}{70 \ \Omega + 50 \ \Omega}$

$$p = \boxed{\tfrac{1}{6}}$$

(b) $Z_{\text{in}} = Z_0 \left(\dfrac{Z_L \cos \beta l + j Z_0 \sin \beta l}{Z_0 \cos \beta l + j Z_L \sin \beta l} \right)$

$$= (50 \ \Omega) \left(\dfrac{70 \cos \dfrac{\pi}{2} + j 50 \sin \dfrac{\pi}{2}}{50 \cos \dfrac{\pi}{2} + j 70 \sin \dfrac{\pi}{2}} \right)$$

$$= \boxed{35.71 \ \Omega}$$

(c) $\text{VSWR} = \dfrac{1 + |\Gamma|}{1 - |\Gamma|} = \dfrac{1 + \frac{1}{6}}{1 - \frac{1}{6}} = \boxed{1.40}$

2. (a) $\dfrac{Z_L}{Z_0} = \dfrac{25 - j25 \ \Omega}{50 \ \Omega} = \tfrac{1}{2} - j\tfrac{1}{2}$ pu

The load point is shown on the Smith chart. The impedance locus is shown as a dashed circle.

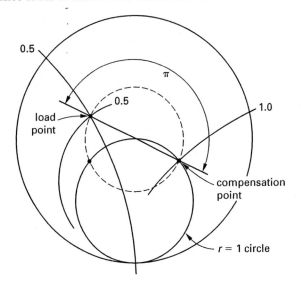

The compensation point for series capacitance is where the impedance locus intersects the $r = 1$ circle on the right-hand side of the chart.

For this particular load, the compensation point is 180° (on the chart) from the load point, at

$$\frac{Z_{in}}{Z_0} = 1 + j1 \text{ pu}$$

180° on the Smith chart is $\lambda/4$, or $\beta l = \boxed{\pi/2 \text{ rad}}$

(b)
$$\frac{Z_{in}}{Z_0} = \frac{\dfrac{Z_L}{Z_0} \cos \beta l + j \sin \beta l}{\cos \beta l + j \left(\dfrac{Z_L}{Z_0} \right) \sin \beta l}$$

$$= \frac{\dfrac{1}{2} \cos \beta l + j \left(\sin \beta l - \dfrac{1}{2} \cos \beta l \right)}{\cos \beta l + \dfrac{1}{2} \sin \beta l + j \dfrac{1}{2} \sin \beta l}$$

At $\beta l = \pi/2$,

$$\sin \beta l = 1$$
$$\cos \beta l = 0$$

$$\frac{Z_{in}}{Z_0} = \frac{j}{\frac{1}{2} + j\frac{1}{2}} = \frac{2}{\frac{1}{j} + 1} = 1 + j1$$

The compensating reactance must be $-j1$, or

$$X_C = -Z_0 = \boxed{-50 \ \Omega}$$

3. (a) $\quad \dfrac{Z}{Z_0} = \dfrac{-j25 \ \Omega}{50 \ \Omega} = -0.5j \text{ pu}$

The bottom of the chart is open.

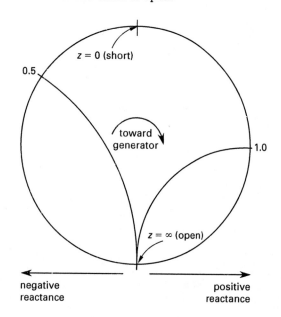

The shortest length to obtain $-j0.5 \times 50$ is to start with an open line, going clockwise 126.5° to the intersection with $x = 0.5$ on the left half of the Smith chart. This corresponds to

$$\beta l = \left(\frac{126.5°}{2} \right) \left(\frac{\pi}{180°} \right) \text{ rad} = 1.1 \text{ rad}$$

$$= (1.1 \text{ rad}) \left(\frac{0.5 \ \lambda}{\pi \text{ rad}} \right)$$

$$= 0.175 \ \lambda$$

(b) $\quad \dfrac{Z}{Z_0} = \dfrac{j50}{50} = j1 \text{ pu}$

The shortest line is found for a shorted line (short point at the top of the Smith chart) going clockwise to intersect the $x = 1$ circle (on the right). This is 90° on the chart, corresponding to

$$\beta l = (45°) \left(\frac{\pi}{180°} \right) \text{ rad} = (0.785 \text{ rad}) \left(\frac{0.5 \ \lambda}{\pi \text{ rad}} \right)$$

$$= \boxed{\lambda/8}$$

4. $\quad Z_0 = 72 \ \Omega$

$\quad Z_L = 50 \ \Omega$

$\quad \beta = 0.5$

$\quad l = 7.5 \text{ m}$

$\quad \dfrac{Z_L}{Z_0} = z_L = \dfrac{50 \ \Omega}{72 \ \Omega} = 0.6944 \text{ pu}$

$\quad \beta l = (3.75 \text{ rad}) \left(\dfrac{360°}{2\pi \text{ rad}} \right) = 214.86°$

$\quad 2\beta l = (2)(214.86°) = 429.72° = 360° + 69.72°$

The load point and source point are shown.

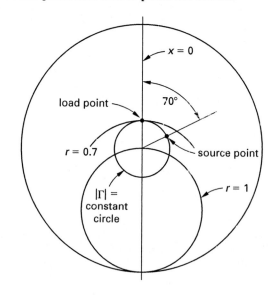

Reading from an enlarged Smith chart, the value is

$$\frac{Z_{in}}{Z_0} \approx 0.84 + j0.29 \text{ pu}$$

$$Z_{in} \approx (0.84 + j0.29 \text{ pu})(72 \text{ } \Omega)$$

$$= \boxed{60 + j21}$$

5. $$Z_0 = 100 \text{ } \Omega$$

$$Z_L = 25 + j25 \text{ } \Omega$$

$$z_L = 0.25 + j0.25$$

$$y_L = 2 - j2$$

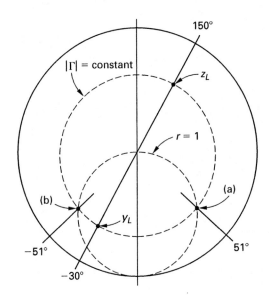

(a) The z_L point, $0.25 + j0.25$, is at $150°$. Following a constant $|\Gamma|$ circle toward the generator, the compensation point is at $51°$, where $z_{in} = 1 + j1.6$. A series compensation of $-j1.6$ can be added. The line length from the load is

$$\left(\frac{150° - 51°}{720°}\right) \lambda = \boxed{0.1375\lambda}$$

$$Z_C = (100)(-j1.6) = -j160 \text{ } \Omega$$

Using a 72 Ω line for the compensator, its reactance is

$$x_C = -\frac{j160 \text{ } \Omega}{72 \text{ } \Omega} = -2.22 \text{ pu}$$

Starting from the open position (bottom of the Smith chart), the location of $-j2.22$ is about $-48°$ for a compensator length of

$$l_{cp} = \left(\frac{48°}{720°}\right) \lambda \approx \boxed{\frac{\lambda}{15}}$$

(b) The y_L point is $180°$ from the z_L point on the same $|\Gamma| = $ constant circle.

The parallel compensation point is at $-51°$, so the distance from the load for compensation is at

$$\left(\frac{51° - 30°}{720°}\right) \lambda = \boxed{0.0292\lambda}$$

At that point,

$$y_L = 1 - j1.6 \text{ pu}$$

$$y = \frac{Z_0}{Z} = Z_0 Y$$

$$Y_{cp} = \frac{y}{Z_0} = j\left(\frac{1.6}{100 \text{ } \Omega}\right) \text{ S}$$

For the 72 Ω line,

$$y_{cp} = j\left(\frac{1.6}{100 \text{ } \Omega}\right)(72 \text{ } \Omega) = j1.152 \text{ pu}$$

1.152 is on the right side of the chart at $82°$. The shortest line is from the top of the chart ($y = 0$, open) at $180°$.

$$l = \left(\frac{180° - 82°}{720°}\right) \lambda = \boxed{0.136\lambda}$$

58 Antenna Theory

PRACTICE PROBLEMS

1. What is the term for a group of components arranged to vary the transmission or reception of electromagnetic waves?
- (A) array
- (B) Huygens' source
- (C) waveguide
- (D) dish antenna

2. For an antenna with a bandwidth of 1 GHz at its maximum design temperature of 65°C, what is the noise power?
- (A) 9×10^{-13} W
- (B) 2×10^{-12} W
- (C) 5×10^{-12} W
- (D) 5×10^{-9} W

3. What is the phase constant for a 500 MHz signal?
- (A) 1.05 rad/m
- (B) 10.5 rad/m
- (C) 32.0 rad/m
- (D) 38.0 rad/m

4. The peak electric field of a plane electromagnetic wave 10 km south of a vertical antenna is 0.2×10^{-2} V/cm. What is the magnitude of the magnetic field at that point?
- (A) 10.0 nA/m
- (B) 1.00 μA/m
- (C) 0.50 mA/m
- (D) 0.05 mA/m

5. A radio station's assigned frequency is 107 MHz. What is the free-space basic transmission loss 50 km from the transmitting antenna?
- (A) 3.5 dB
- (B) 47 dB
- (C) 83 dB
- (D) 110 dB

SOLUTIONS

1. An array is used to manipulate the direction of an antenna with respect to incoming or outgoing electromagnetic waves.

The answer is (A).

2. The noise power is

$$P_n = \kappa TB$$

The design temperature of 65°C equates to 338K. Substituting gives

$$P_n = \left(1.3805 \times 10^{-23} \; \frac{\text{J}}{\text{K}}\right)(338\text{K})(10^9 \; \text{Hz})$$

$$= \boxed{4.67 \times 10^{-12} \; \text{W} \quad (5 \times 10^{-12} \; \text{W})}$$

The answer is (C).

3. The phase constant, β, is

$$\beta = \frac{2\pi}{\lambda}$$

The wavelength of a 500 MHz wave is

$$\lambda = \frac{c}{f} = \frac{3 \times 10^8 \; \frac{\text{m}}{\text{s}}}{500 \times 10^6 \; \frac{1}{\text{s}}}$$

$$= 0.6 \; \text{m}$$

Substituting gives

$$\beta = \frac{2\pi}{\lambda} = \frac{2\pi}{0.6 \; \text{m}} = \boxed{10.5 \; \text{rad/m}}$$

The answer is (B).

4. From Poynting's vector,

$$\mathbf{S} = \mathbf{E} \times \mathbf{H} = \epsilon c E^2 \mathbf{a}$$

Therefore,

$$|\mathbf{S}| = EH = \epsilon c E^2$$

Air has a relative permittivity, ϵ_r, of approximately one. Thus, $\epsilon = \epsilon_0 \epsilon_r \approx \epsilon_0$. Rearranging and substituting gives

$$EH = \epsilon c E^2$$

$$H = \epsilon c E \approx \epsilon_0 c E$$

$$= \left(8.854 \times 10^{-12} \ \frac{\text{F}}{\text{m}}\right)\left(3 \times 10^8 \ \frac{\text{m}}{\text{s}}\right)$$

$$\times \left(0.2 \times 10^{-2} \ \frac{\text{V}}{\text{cm}}\right)\left(100 \ \frac{\text{cm}}{\text{m}}\right)$$

$$= 5.31 \times 10^{-4} \ \text{F·V/m·s}$$

$$= \boxed{5.31 \times 10^{-4} \ \text{A/m} \quad (0.5 \ \text{mA/m})}$$

The answer is (C).

5. The free-space basic transmission loss is

$$L_{b0} = 20 \log \frac{4\pi r}{\lambda}$$

The wavelength is determined from

$$\lambda = \frac{c}{f} = \frac{3 \times 10^8 \ \frac{\text{m}}{\text{s}}}{107 \times 10^6 \ \frac{1}{\text{s}}} = 2.8 \ \text{m}$$

Substituting gives

$$L_{b0} = 20 \log \frac{4\pi(50 \times 10^3 \ \text{m})}{2.8 \ \text{m}}$$

$$= \boxed{107 \ \text{dB} \quad (110 \ \text{dB})}$$

The answer is (D).

59 Communication Links

PRACTICE PROBLEMS

1. A radio station transmits 50 kW with a directivity gain of 1.5. What is the power density 10 km from the radiating antenna?
- (A) 0.060 mW/m^2
- (B) 5.9 W/m^2
- (C) 33 W/m^2
- (D) 75 W/m^2

2. For the radio station in Prob. 1, what is the electric field strength at 10 km from the radiating antenna?
- (A) 0.080 V/m
- (B) 0.15 V/m
- (C) 50 V/m
- (D) 100 V/m

3. An antenna has an effective aperture of 10 m^2. What additional information is required to determine the amplifier power, P_{amp}?
- (A) power density
- (B) power loss
- (C) space loss
- (D) total aperture

4. A horn antenna with an efficiency of 90% has a directivity of 25. What is the gain in decibels?
- (A) 1.0 dB
- (B) 14 dB
- (C) 23 dB
- (D) 28 dB

5. Two antennae used for wireless communication at 900 MHz are identical, with physical apertures of 0.1 m^2 and a total efficiency of 50%. The transmitting power is 1 W. The antennae provide adequate reception if the received power density is 10 μW. What is the maximum separation distance?
- (A) 15 m
- (B) 25 m
- (C) 50 m
- (D) 150 m

SOLUTIONS

1. The power density is

$$p_r = \left(\frac{D_T}{4\pi r^2}\right) P_T$$

Substituting gives

$$p_r = \left(\frac{1.5}{4\pi(10 \times 10^3 \text{ m})^2}\right)(50 \times 10^3 \text{ W})$$

$$= \boxed{5.97 \times 10^{-5} \text{ W/m}^2 \quad (0.06 \text{ mW/m}^2)}$$

The answer is (A).

2. The power density in terms of the radiating electric field is

$$p_r = \frac{E^2}{Z_0}$$

The characteristic impedance of the atmosphere is approximately that of free space. Thus, rearranging gives

$$E = \sqrt{p_r Z_0}$$

$$= \sqrt{\left(0.06 \times 10^{-3} \frac{\text{W}}{\text{m}^2}\right)(120\pi \text{ }\Omega)}$$

$$= \boxed{0.15 \text{ V/m}}$$

The answer is (B).

3. The amplifier power is

$$P_{amp} = A_{eff}p_r$$

Thus, the information required is the power density, p_r.

The answer is (A).

4. The gain is

$$G = \eta D_g$$

Substituting the given value gives

$$G = (0.90)(25) = 22.5$$

In decibels, the gain is

$$G_{dB} = 10 \log 22.5$$

$$= \boxed{13.5 \text{ dB} \quad (14 \text{ dB})}$$

The answer is (B).

5. The transmission equation is

$$P_R = \left(\frac{\lambda}{4\pi r}\right)^2 G_R G_T P_T$$

The gains are unknown. The gains are given by

$$G = \eta D$$

Substituting gives

$$P_R = \left(\frac{\lambda}{4\pi r}\right)^2 \eta_R D_R \eta_T D_T P_T$$

The two antennae are identical. Thus,

$$P_R = \left(\frac{\lambda}{4\pi r}\right)^2 (\eta D)^2 P_T$$

The directivity is unknown. The directivity is given by

$$D = \left(\frac{4\pi}{\lambda^2}\right) A_{eff} = \left(\frac{4\pi}{\lambda^2}\right) A_{physical}$$

The physical aperture can be used because the total efficiency is given, which includes the antenna $I^2 R$ efficiencies and is accounted for in ηD. Alternatively, the gain is

$$G = \left(\frac{4\pi}{\lambda^2}\right) \eta_{total} A_{physical}$$

Substituting the directivity D into P_R gives

$$P_R = \left(\frac{\lambda}{4\pi r}\right)^2 \left(\eta_{total}\left(\frac{4\pi}{\lambda^2}\right) A_{physical}\right)^2 P_T$$

$$= \left(\frac{\eta_{total}^2}{\lambda^2 r^2}\right) A_{physical}^2 P_T$$

Now, the only unknown is the radial distance r. Rearranging to solve for r gives

$$r = \sqrt{\frac{\eta_{total}^2 A_{physical}^2 P_T}{P_R \lambda^2}}$$

Substituting the definition of the wavelength and the given values results in

$$r = \sqrt{\frac{\eta_{total}^2 A_{physical}^2 P_T}{P_R \left(\frac{c}{f}\right)^2}}$$

$$= \left(\frac{\eta_{total} A_{physical}}{\frac{c}{f}}\right)\sqrt{\frac{P_T}{P_R}}$$

$$= \left(\frac{(0.5)(0.1 \text{ m}^2)}{\dfrac{3 \times 10^8 \, \frac{m}{s}}{900 \times 10^6 \, \frac{1}{s}}}\right)\sqrt{\frac{1 \text{ W}}{10 \times 10^{-6} \text{ W}}}$$

$$= \boxed{47.4 \text{ m} \quad (50 \text{ m})}$$

The answer is (C).

60 Signal Formats

PRACTICE PROBLEMS

1. Commercial AM radio generally uses what type of modulation?
- (A) DSB-SC
- (B) DSB-LC
- (C) QAM
- (D) VSAM

2. The maximum frequency deviation for U.S. AM radio stations is 5 kHz. The allowable frequency band is 535 kHz to 1705 kHz. How many radio stations can a single geographical area support?
- (A) 117 stations
- (B) 170 stations
- (C) 234 stations
- (D) 341 stations

3. A 105 MHz carrier is frequency modulated by a 75 kHz information signal. The information signal has a 1 V amplitude and a frequency modulator constant of 100 Hz/V. What is the bandwidth?
- (A) 75 kHz
- (B) 100 kHz
- (C) 150 kHz
- (D) 200 kHz

4. The same carrier in Prob. 3 is now phase modulated, with $k_p = 0.3$ rad/V. What is the bandwidth?
- (A) 75 kHz
- (B) 100 kHz
- (C) 150 kHz
- (D) 200 kHz

5. A communication system design calls for multilevel modulation of a digital signal. A mapping space using four-digit binary codes has been selected. How many potential modulations of the carrier are there?
- (A) 2 modulations
- (B) 4 modulations
- (C) 8 modulations
- (D) 16 modulations

6. For the NRZ-M baseband data signal shown, what binary signal is transmitted, assuming that the most significant bit is transmitted first?

- (A) 0010
- (B) 0110
- (C) 1001
- (D) 1101

7. If the signal in Prob. 6 is inverted, what data is sent?
- (A) 0010
- (B) 0110
- (C) 1001
- (D) 1101

8. What is the bandwidth efficiency of a one-dimensional 4-ary PAM signal?
- (A) 1
- (B) 2
- (C) 4
- (D) 8

9. If the bandwidth in Prob. 8 is 10 kHz, what is the signal rate or bit rate in kilobits per second?
- (A) 5 kbps
- (B) 10 kbps
- (C) 40 kbps
- (D) 80 kbps

10. Intersymbol interference is eliminated by applying which of the following?
- (A) Nyquist criterion
- (B) pulse code modulation
- (C) Shannon bandwidth
- (D) SSB modulation

11. What is the probability of error for a 4-ary PAM scheme, assuming $E_b/N_0 = 1$?

(A) 0.15
(B) 0.30
(C) 0.60
(D) 0.90

12. What is the bit error rate of the 4-ary PAM scheme in Prob. 11?

(A) 0.15
(B) 0.30
(C) 0.60
(D) 0.90

SOLUTIONS

1. Commercial AM radio uses a double-sideband large-carrier (DSB-LC) signal, because this signal utilizes simple, inexpensive receivers.

The answer is (B).

2. The station bandwidth is given by

$$BW = 2\Delta f$$

$$= (2)\left(5 \; \frac{kHz}{station}\right)$$

$$= 10 \; \frac{kHz}{station}$$

The AM bandwith is

$$AM \; BW = 1705 \; kHz - 535 \; kHz = 1170 \; kHz$$

The number of possible stations is

$$stations = \frac{AM \; BW}{station \; BW} = \frac{1170 \; kHz}{\dfrac{10 \; kHz}{station}}$$

$$= \boxed{117 \; stations}$$

The answer is (A).

3. The bandwidth of an FM signal is approximated using Carson's rule.

$$BW \approx 2(\Delta\omega + \omega_{mod})$$

The maximum frequency deviation for an FM signal is

$$\Delta\omega = ak_f$$

Substituting the given values results in

$$BW \approx 2(\Delta\omega + \omega_{mod}) = 2(ak_f + \omega_{mod})$$

$$= (2)\left((1 \; V)\left(100 \; \frac{Hz}{V}\right) + 75 \times 10^3 \; Hz\right)$$

$$= \boxed{150{,}200 \; Hz \quad (150 \; kHz)}$$

The answer is (C).

4. For a phase-modulated signal, Carson's rule still applies, but the maximum frequency deviation is

$$\Delta\omega = ak_p\omega_{mod} = \Delta\theta\omega_{mod}$$

$$= (1 \; V)\left(0.3 \; \frac{rad}{V}\right)(75{,}000 \; Hz)$$

$$= 22{,}500 \; Hz$$

Substituting into Carson's rule gives

$$BW \approx 2(\Delta\omega + \omega_{mod})$$

$$= 2(\Delta\theta\omega_{mod} + \omega_{mod})$$

$$= (2)(22{,}500 \text{ Hz} + 75{,}000 \text{ Hz})$$

$$= \boxed{195{,}000 \text{ Hz} \quad (200 \text{ kHz})}$$

The answer is (D).

5. The number of mapping locations, M, for binary digits, n, is given by

$$M = 2^n = (2)^4 = 16$$

There are 16 potential modulations of the carrier.

The answer is (D).

6.

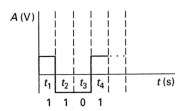

1 1 0 1

Assuming that the signal starts at zero, the first interval represents a change and thus equals a logic 1.

A change occurs in the consecutive signals from bit interval t_1 to t_2, and thus the second interval equals logic 1.

No change occurs from t_2 to t_3, and thus t_3 is logic 0.

A change occurs from t_3 to t_4, and thus t_4 is logic 1. The net result is a signal of 1101.

The answer is (D).

7. Inverting an NRZ-M signal does not change the transitions from one bit interval to the next. Thus, the logic circuitry is able to detect the same signal, 1101.

The answer is (D).

8. The bandwidth efficiency (noting that $D = 1$) is defined as

$$\eta_{BW} = \frac{R_s}{BW} = \frac{\dfrac{\log_2 M}{T}}{\dfrac{D}{2T}} = \frac{2\log_2 M}{D}$$

$$= 2\log_2 4$$

$$= \boxed{4}$$

The answer is (C).

9. The bandwidth efficiency is

$$\eta_{BW} = \frac{R_s}{BW}$$

Solving for the signal rate and substituting the given values and the calculated efficiency from Prob. 8 gives

$$R_s = \eta_{BW}(BW)$$

$$= (4)(10 \text{ kHz})$$

$$= \boxed{40 \text{ kHz} \quad (40 \text{ kbps})}$$

The answer is (C).

10. The Nyquist criterion is specifically designed to achieve zero intersymbol interference.

The answer is (A).

11. The probability of error for a PAM scheme is

$$\mathcal{P}(e) = \left(1 - \frac{1}{M}\right)\text{erfc}\sqrt{\left(\frac{3\log_2 M}{M^2 - 1}\right)\left(\frac{E_b}{N_0}\right)}$$

The number of bits in a symbol for an M-ary modulation scheme is

$$\text{bits} = \log_2 M = \log_2 4 = 2$$

The symbols are 00, 01, 10, and 11 in a 4-ary scheme with 2 bits per symbol.

Substituting $M = 4$ and $E_b/N_0 = 1$ gives

$$\mathcal{P}(e) = \left(1 - \frac{1}{4}\right)\text{erfc}\sqrt{\left(\frac{3\log_2 4}{(4)^2 - 1}\right)(1)}$$

$$= \left(\frac{3}{4}\right)\text{erfc}\sqrt{\frac{6}{15}}$$

$$= \left(\frac{3}{4}\right)\text{erfc}\sqrt{\frac{2}{5}}$$

$$= (0.75)\,\text{erfc}\,(0.63)$$

From a table of the error function, also called the probability function or error integral,

$$\text{erfc}(0.63) = 1 - \text{erf}(0.63)$$

$$= 1 - 0.6270$$

$$= 0.373$$

Thus,

$$\mathcal{P}(e) = (0.75)(0.373) = \boxed{0.28 \quad (0.30)}$$

Note: The bit error rate is high because the bit energy, E_b, is low. In this case, the bit energy is equal to the noise, N_0.

The answer is (B).

12. The bit errror rate (BER) is

$$P_b(e) = \frac{P(e)}{\log_2 M}$$

The probability of error for a 4-ary PAM is calculated in Prob. 11. The number of bits is given by the denominator of the BER equation and is calculated in Prob. 11. Substituting gives

$$P_b(e) = \frac{P(e)}{\log_2 M} = \frac{0.28}{2}$$

$$= \boxed{0.14 \quad (0.15)}$$

The answer is (A).

61 Signal Multiplexing

PRACTICE PROBLEMS

1. Which of the following forms of multiplexing is an example of spread-spectrum transmission?

- (A) CDMA
- (B) FDM
- (C) SDM
- (D) TDM

2. Consider the bit-stream illustration shown. What type of multiplexing is represented?

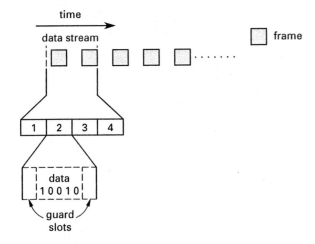

- (A) FDM
- (B) SDM
- (C) TDM
- (D) TDMA

3. In optical communication, frequency-division multiplexing (FDM) is called

- (A) PDM
- (B) SDM
- (C) SSB
- (D) WDM

4. What analysis technique can be used to determine the effect of multiple frequency sources, at different frequencies, operating on the same linear circuit?

- (A) Kirchhoff's laws
- (B) Miller's theorem
- (C) Shannon's theorem
- (D) superposition

5. What form of multiplexing depends on geographical separation of the signal sources?

- (A) FM
- (B) MUX
- (C) SDM
- (D) TDMA

SOLUTIONS

1. Spread-spectrum multiplexing uses a wide bandwidth with power thinly spread over the spectrum—near the noise level. Two forms of spread-spectrum multiplexing are code-division multiple access (CDMA), and time-division multiple access (TDMA).

The answer is (A).

2. The data frames are separated in time with guard slots for synchronization. Thus, time-division multiplexing (TDM) is shown.

The answer is (C).

3. In the optical communications field, FDM is referred to as wavelength-division multiplexing (WDM).

The answer is (D).

4. Superposition may be used to determine the impact of multiple sources acting on the same linear circuit.

The answer is (D).

5. Geographical separation of signals is obtained by lowering the power of adjacent sources, use of spot beams, or different polarizations. All the techniques are a form of space-division multiplexing (SDM).

The answer is (C).

62 Communication Systems and Channels

PRACTICE PROBLEMS

1. A microwave cellular telephone system is built with antenna towers 90 m high. What is most nearly the maximum distance between towers?

(A) 20 km
(B) 30 km
(C) 40 km
(D) 80 km

2. A certain communication channel has a signal-to-noise ratio of 7 and a bandwidth of 10 kHz. What is the channel capacity?

(A) 30 bps
(B) 80 bps
(C) 3000 bps
(D) 30,000 bps

3. A copper cable used in a POTS line has a loss of 1.5 dB/km. What is the approximate loss experienced if the user is located 15 km from the central office?

(A) 1.5 dB
(B) 15 dB
(C) 20 dB
(D) 220 dB

4. A superheterodyne AM receiver with an intermediate frequency (IF) of 455 kHz is to be tuned to listen to a station at 540 kHz. What tuning frequency is used in the local oscillator?

(A) 95 kHz
(B) 540 kHz
(C) 550 kHz
(D) 995 kHz

5. When the receiver in Prob. 4 is tuned to the 540 kHz station, what other AM station offers potential interference?

(A) 85 kHz
(B) 995 kHz
(C) 1080 kHz
(D) 1450 kHz

SOLUTIONS

1. Microwave signals are line-of-sight signals. The radio horizon for an antenna is

$$d = 4.1\sqrt{h}$$

The maximum distance between two towers, d, is

$$d = \left((4.1)\sqrt{90 \text{ m}}\right)(2)$$

$$= \boxed{77.8 \text{ km} \quad (80 \text{ km})}$$

The situation is shown.

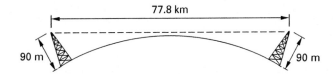

The answer is (D).

2. The channel capacity is

$$C = (\text{BW}) \ \log_2\left(1 + \frac{S}{N}\right)$$

Substitute the given information.

$$C = (10 \times 10^3 \text{ Hz}) \ \log_2(1 + 7)$$

$$= \boxed{30,000 \text{ bps}}$$

The answer is (D).

3. The given loss is 1.5 dB/km. The total loss is

$$L_{\text{dB}} = \left(1.5 \ \frac{\text{dB}}{\text{km}}\right)(15 \text{ km})$$

$$= \boxed{22.5 \text{ dB} \quad (20 \text{ dB})}$$

The answer is (C).

4. The process of heterodyning mixes the desired signal with the intermediate frequency (IF), generating the sum and difference. Thus, the heterodyned frequencies are

$$f_H = \text{signal} \pm \text{IF}$$

$$= 540 \text{ kHz} \pm 455 \text{ kHz}$$

$$= \boxed{995 \text{ kHz}} \text{ and } 85 \text{ kHz}$$

The highest of the two is always used, leading to a proportionally smaller band over which the oscillator must tune.

The answer is (D).

5. The oscillator frequency in Prob. 4 is 995 kHz, which is also the difference frequency between a 1450 kHz station and a 455 kHz IF. This station, called the *image station*, is located at 2(IF) from the desired station. IF filters remove the image station signal of 1450 kHz.

The answer is (D).

63 Biomedical Electrical Engineering

PRACTICE PROBLEMS

The following information is applicable to Probs. 1 and 2.

A medical incubator is designed to maintain the temperature, humidity, and CO_2 at the proper levels for the sampled items. The humidifier uses the low-water-level alarm circuitry shown.

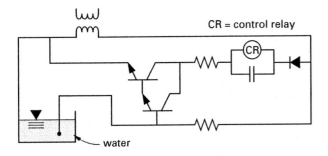

1. What term describes the configuration of the two transistors?

(A) common collector

(B) common base

(C) common source

(D) Darlington pair

2. Which of the following transistor parameters represents the gain used to drive the control relay?

(A) h_{fe}

(B) h_{ie}

(C) h_{oe}

(D) h_{re}

3. A Class II biological safety cabinet (hood) uses a 115 V single-phase AC motor to maintain the vacuum level within the hood necessary to prevent the spread of biological contaminants. With clean filters, the motor draws 10 A at its rated power factor of 0.85 lagging. What is the power consumption of the motor?

(A) 980 W

(B) 1150 W

(C) 1700 W

(D) 1990 W

4. Leakage current standards for biomedical devices are generally in the microamp (μA) range. Which of the following items is (are) used to reduce leakage currents for 60 Hz systems?

(A) DC-DC inverters

(B) rechargeable batteries

(C) motor-generators

(D) all of the above

5. Radiofrequency units used for patient care may result in _____ human tissue.

(A) reorientation of

(B) heating of

(C) increased circulation within

(D) reorganization of cellular structure

SOLUTIONS

1. The two transistors are in a common emitter configuration (analogous to a common source configuration for field-effect transistors (FETs)). This specific configuration is more commonly referred to as a Darlington pair.

The answer is (D).

2. The Darlington pair has a high input resistance and a high current gain (h_{fe}). It is this forward current gain that enables operation of the control relay.

The answer is (A).

3. The power in a single-phase motor is

$$P = IV(\text{pf})$$

$$= (10 \text{ A})(115 \text{ V})(0.85)$$

$$= \boxed{977.5 \text{ W} \quad (980 \text{ W})}$$

The answer is (A).

4. Leakage currents are minimized by using nonoscillatory power sources (such as DC-DC inverters and rechargeable batteries) or electrical items with nonelectromagnetic coupling (such as motor-generators).

The answer is (D).

5. Radiofrequency fields of sufficient intensity can damage human cells by deposition of energy through heating.

The answer is (B).

64 Analysis of Engineering Systems: Control Systems

PRACTICE PROBLEMS

1. Consider a homogeneous second-order linear differential equation with constant coefficients. The damping ratio is 0.3, and the natural frequency is 10 rad/s. Estimate the amplitude at $t = 0.5$ s graphically.

2. Determine the transfer functions for the following systems.

(a)

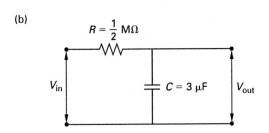

(b)

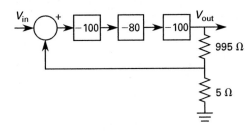

3. An amplifier consists of several stages with gains as shown. A voltage divider provides feedback. What is the overall gain with and without feedback?

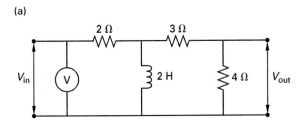

4. A system is composed of a forward path and a unity feedback loop. The gain in the forward loop is 1000 $\pm10\%$. What is the uncertainty of the output signal?

5. (a) Simplify the following block diagrams and (b) determine the overall system gain.

(a)

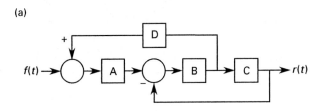

(b)

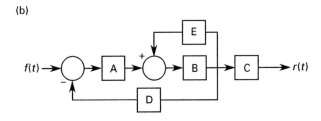

6. (a) Simplify the following block diagram and (b) determine the overall system gain. (c) What is the system sensitivity if $G_1 = -5$, $G_2 = 2$, $G_3 = 4$, and $G_4 = 3$?

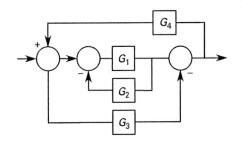

7. What is the steady-state response to an impulse $\delta(t)$ of the system represented by the following differential equation? Assume all initial conditions are zero.

$$r''(t) + 3r'(t) + r(t) = \delta'(t) + \delta(t)$$

8. Find the steady-state response to a step input for a system in which the transfer function is

$$T(s) = \frac{b_0 s^p + b_1 s^{p-1} + \cdots + b_p}{s^n + a_1 s^{n-1} + \cdots + a_n} \quad [a_n \neq 0]$$

9. Consider the following transfer function.

$$T(s) = \frac{1+s}{1+2s+2s^2}$$

What is the (a) amplitude and (b) phase of the steady-state response to a sinusoidal input of $\sin \omega t$?

10. Consider a system with the following transfer function.

$$T(s) = \frac{1}{(s+a)(s+b)}$$

What is the steady-state response to a unit step (a) in the time domain and (b) in the frequency domain?

11. Draw the pole-zero diagram for the following transfer function.

$$T(s) = \frac{(s^2+4)(s-2)}{s^2(s^2+4s+5)(s+1)}$$

12. What is the steady-state response of a system to a sinusoidal input with angular frequency ω_0 if the system's transfer function has a zero at $s = j\omega_0$?

13. Find the (a) bandwidth, (b) peak frequency, (c) half-power points, and (d) quality factor for

$$T(s) = \frac{3s+18}{s^2+12s+3200}$$

14. Is a system with the following transfer function stable?

$$T(s) = \frac{(s-2)(s+3)}{(s^2+s-2)(s+1)}$$

SOLUTIONS

1. At $t = 0.5$ s,

$$\omega t = \left(10 \; \frac{\text{rad}}{\text{s}}\right)(0.5 \text{ s}) = 5 \text{ rad}$$

For a normalized curve of natural response corresponding to $\zeta = 0.3$, see Fig. 64.1.

$$\frac{x(t)}{\omega} \approx -0.22 \text{ s/rad} \quad [\text{at } t = 0.5]$$

$$x(t = 0.5 \text{ s}) = \left(-0.22 \; \frac{\text{s}}{\text{rad}}\right)\left(10 \; \frac{\text{rad}}{\text{s}}\right)$$

$$= \boxed{-2.2}$$

2. (a)

At node 2,

$$0 = \frac{V_2 - V_1}{R_1} + \frac{V_2 - V_3}{R_2} + \frac{1}{L}\int V_2 dt$$

$$= \frac{V_2 - V_1}{2} + \frac{V_2 - V_3}{3} + \frac{1}{2}\int V_2 dt$$

$$= \frac{V_2 - V_1}{2} + \frac{V_2 - V_3}{3} + \frac{V_2}{2s} \qquad [\text{Eq. I}]$$

At node 3,

$$0 = \frac{V_3 - V_2}{R_2} + \frac{V_3}{R_3}$$

$$= \frac{V_3 - V_2}{3} + \frac{V_3}{4} \qquad [\text{Eq. II}]$$

From Eq. I,

$$V_2 = \frac{3sV_1 + 2sV_3}{5s+3}$$

From Eq. II,

$$V_2 = \frac{7V_3}{4}$$

$$\frac{3sV_1 + 2sV_3}{5s+3} = \frac{7V_3}{4}$$

$$T(s) = \frac{V_\text{out}}{V_\text{in}} = \frac{V_3}{V_1} = \frac{12s}{27s+21} = \boxed{\frac{4s}{9s+7}}$$

The black box representation of this system is

$$V_\text{in} \rightarrow \boxed{\dfrac{4s}{9s+7}} \rightarrow V_\text{out}$$

(b)

$$0 = \frac{V_\text{out} - V_\text{in}}{R} + C\left(\frac{dV_\text{out}}{dt}\right)$$

$$0 = \frac{V_\text{out} - V_\text{in}}{R} + CsV_\text{out}$$

$$0 = V_\text{out} - V_\text{in} + CRsV_\text{out}$$

$$T(s) = \frac{V_\text{out}}{V_\text{in}} = \frac{1}{1 + CRs}$$

$$= \frac{1}{1 + (3 \times 10^{-6}\ \text{F})\left(\dfrac{1}{2} \times 10^6\ \Omega\right) s}$$

$$= \boxed{\frac{2}{3s+2}}$$

$$V_\text{in} \rightarrow \boxed{\dfrac{2}{3s+2}} \rightarrow V_\text{out}$$

3.

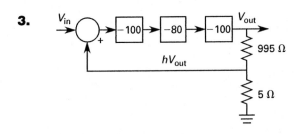

The overall gain without feedback is

$$K = (-100)(-80)(-100) = \boxed{-800{,}000}$$

$$\frac{V_\text{out} - hV_\text{out}}{995\ \Omega} = \frac{hV_\text{out}}{5\ \Omega}$$

$$h = \frac{5\ \Omega}{995\ \Omega + 5\ \Omega} = 0.005$$

$K < 0$, but the feedback adds, so the feedback is negative.

$$K_\text{loop} = \frac{K}{1 - Kh} = \frac{-800{,}000}{1 - (-800{,}000)(0.005)} = \boxed{-200}$$

4.

$$v_\text{out} = G_\text{loop} v_i$$

$$\frac{\Delta v_\text{out}}{v_\text{out}} = \frac{\Delta G_\text{loop}}{G_\text{loop}}$$

(Assume the input signal has no uncertainty.)

$$\left(\frac{\Delta G_\text{loop}}{G_\text{loop}}\right)\left(\frac{G}{\Delta G}\right) = \frac{1}{1+G}$$

$$\frac{\Delta v_\text{out}}{v_\text{out}} = \frac{\Delta G_\text{loop}}{G_\text{loop}} = \left(\frac{\Delta G}{G}\right)\left(\frac{1}{1+G}\right)$$

$$\frac{\Delta G}{G} = 0.1$$

$$\frac{\Delta v_\text{out}}{v_\text{out}} = (0.1)\left(\frac{1}{1+1000}\right) = 9.99 \times 10^{-5}$$

$$\approx \boxed{0.01\%}$$

5. (a)

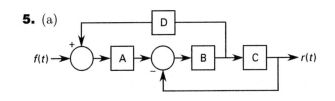

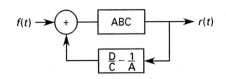

$$f(t) \rightarrow \boxed{\dfrac{ABC}{1 - ABD + BC}} \rightarrow r(t)$$

$$\text{overall system gain} = \frac{r(t)}{f(t)} = \boxed{\frac{ABC}{1 - ABD + BC}}$$

(b)

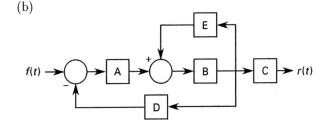

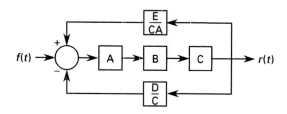

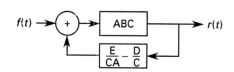

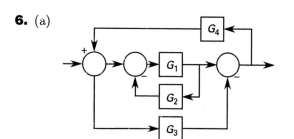

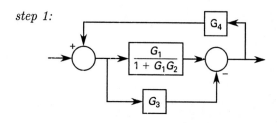

overall system gain $= \dfrac{r(t)}{f(t)} = \boxed{\dfrac{ABC}{1 - BE + ABD}}$

6. (a)

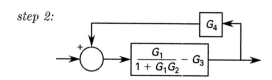

step 1:

step 2:

step 3:

(b) The overall system gain is

$$\boxed{\dfrac{G_1 - G_3 - G_1 G_2 G_3}{1 + G_1 G_2 - G_1 G_4 + G_3 G_4 + G_1 G_2 G_3 G_4}}$$

From step 2,

$$G(s) = \dfrac{G_1}{1 + G_1 G_2} - G_3$$

$$= \dfrac{-5}{1 + (-5)(2)} - 4$$

$$= -31/9$$

$$H(s) = G_4 = 3$$

$$G(s)H(s) = \left(-\dfrac{31}{9}\right)(3) = -\dfrac{31}{3} < 0$$

(c) The system has negative feedback overall. Therefore,

$$\text{sensitivity } S = \dfrac{1}{1 + G(s)H(s)}$$

$$= \dfrac{1}{1 + \left(-\dfrac{31}{3}\right)}$$

$$= \boxed{-3/28}$$

7. By definition, the system transfer function is

$$T(s) = \mathcal{L}\left(\dfrac{r_{\text{sys}}(t)}{f_{\text{sys}}(t)}\right) = \dfrac{R_{\text{sys}}(s)}{F_{\text{sys}}(s)}$$

$$= \dfrac{\text{output signal}}{\text{input signal}}$$

The output signal is

$$r_{\text{sys}}(t) = \delta'(t) + \delta(t)$$

$$R_{\text{sys}}(s) = s + 1$$

The input signal is

$$f_{\text{sys}}(t) = r''(t) + 3r'(t) + r(t)$$

$$F_{\text{sys}}(s) = s^2 + 3s + 1$$

$$T(s) = \dfrac{R_{\text{sys}}(s)}{F_{\text{sys}}(s)} = \dfrac{s + 1}{s^2 + 3s + 1}$$

The forcing function, $f(t)$, acting on this system is

$$f(t) = \delta(t)$$

$$F(s) = \mathcal{L}(f(t)) = \mathcal{L}(\delta(t)) = 1$$

$$R(s) = T(s)F(s) = (T(s))(1)$$

$$= T(s)$$

$$= \boxed{\dfrac{s + 1}{s^2 + 3s + 1}}$$

8. Using the final value theorem, obtain the steady-state step response by substituting zero for s in the transfer function.

If $b_p \neq 0$,

$$\boxed{R(s) = T(0) = b_p/a_n}$$

If $b_p = 0$, the numerator is zero.

$$\boxed{R(s) = 0}$$

9. (a) Substitute $j\omega$ for s in the transfer function to obtain the steady-state response for a sinusoidal input.

$$R(s) = T(j\omega) = \frac{1 + j\omega}{1 + 2j\omega - 2\omega^2} = \frac{1 + j\omega}{(1 - 2\omega^2) + 2j\omega}$$

The amplitude is the absolute value of $R(s)$.

$$|R(s)| = \sqrt{T^2(j\omega)} = \frac{\sqrt{1 + \omega^2}}{\sqrt{(1 - 2\omega^2)^2 + 4\omega^2}}$$

$$= \boxed{\sqrt{\frac{1 + \omega^2}{1 + 4\omega^4}}}$$

(b) The phase angle of the steady-state response is $\text{Arg}(R(s))$. Find this by substituting $T(j\omega)$ into the form $(a + jb)/c$, whose argument (Arg) is

$$\arctan \frac{\frac{b}{c}}{\frac{a}{c}} = \arctan \frac{b}{a}$$

First eliminate j from the denominator by multiplying by its complex conjugate $(1 - 2\omega^2) - 2j\omega$.

$$R(s) = T(j\omega) = \frac{(1 + j\omega)(1 - 2\omega^2 - 2j\omega)}{1 + 4\omega^4}$$

$$= \frac{1 - 2\omega^2 + 2\omega^2 + j(-\omega - 2\omega^3)}{1 + 4\omega^4}$$

$$= \frac{1 + j(-\omega - 2\omega^3)}{1 + 4\omega^4}$$

$$\text{Arg}(R(s)) = \text{Arg}(T(j\omega))$$

$$= \boxed{\arctan \frac{-(\omega + 2\omega^3)}{1}}$$

10. (a) $R(s) = T(s)F(s)$

$$f(t) = \mu_0 \quad \text{[unit step]}$$

$$F(s) = \mathcal{L}(f(t)) = 1/s$$

$$R(s) = T(s)\left(\frac{1}{s}\right) = \frac{1}{s(s+a)(s+b)}$$

From the transform table,

$$r(t) = \mathcal{L}^{-1}(R(s)) = \frac{1}{ab} + \frac{be^{-at} - ae^{-bt}}{ab(a - b)}$$

$$= r_1 + r_2(t)$$

The steady-state response is $\lim_{t\to\infty} r(t)$, and because $\lim_{t\to\infty} r_2(t) = 0$, r_1 is the steady-state response.

$$r_1 = \boxed{1/ab}$$

(b) The steady-state response in the frequency domain for a step input of magnitude h is $R(s) = hT(0)$. Substituting $h = 1$ and $s = 0$ in $T(s)$,

$$R(s) = h(T(0)) = \frac{1}{(0 + a)(0 + b)} = \boxed{1/ab}$$

11.

$$\frac{(s^2 + 4)(s - 2)}{s^2(s^2 + 4s + 5)(s + 1)} = \frac{(s - 2j)(s + 2j)(s - 2)}{s^2(s + 1)(s + 2 - j)(s + 2 + j)}$$

For zeros, set the numerator equal to zero. For poles, set the denominator equal to zero.

12. The amplitude and phase of the steady-state response are given by $T(j\omega_o)$, where $T(s)$ is the transfer function. Since $j\omega_o$ is a zero of $T(s)$,

$$R(s) = T(j\omega_o) = 0$$

Therefore, the steady-state response is zero.

The system entirely blocks the angular frequency, ω_o.

13. (a)

$$T(s) = \frac{3s + 18}{s^2 + 12s + 3200}$$

$$= \frac{as + b}{s^2 + (BW)s + \omega_n{}^2}$$

$$\text{bandwidth} = BW = \boxed{12 \text{ rad/s}}$$

(b)

$$\text{peak frequency} = \omega_n$$

$$= \sqrt{3200}$$

$$= \boxed{56.57 \text{ rad/s}}$$

(c)

$$\text{half-power points} = \omega_n \pm \frac{BW}{2}$$

$$= 56.57 \, \frac{\text{rad}}{\text{s}} \pm \frac{12 \, \frac{\text{rad}}{\text{s}}}{2}$$

$$= \boxed{\begin{array}{l} 62.57 \text{ rad/s and} \\ 50.57 \text{ rad/s} \end{array}}$$

(d)

$$\text{quality factor} = Q = \frac{\omega_n}{BW}$$

$$= \frac{56.57 \, \frac{\text{rad}}{\text{s}}}{12 \, \frac{\text{rad}}{\text{s}}} = \boxed{4.71}$$

14. The system has a pole in the right half of the s-plane ($s = 1$).

$$s^2 + s - 2 = (s - 1)(s + 2)$$

$$\boxed{\text{The system is unstable.}}$$

65 Electrical Materials

PRACTICE PROBLEMS

1. Which of the following types of material has high thermal and electrical conductivity?

(A) ceramic
(B) metal
(C) polymer
(D) semiconductor

2. Which of the following elements is a liquid at standard atmospheric temperature and pressure (STP)?

(A) Ge
(B) H
(C) Hg
(D) In

3. Which of the following materials has the highest conductivity at room temperature?

(A) aluminum
(B) copper
(C) gold
(D) silver

4. Which of the following terms does NOT describe a ceramic material?

(A) crystalline
(B) high electrical resistance
(C) high electrical conductivity
(D) insulator

5. Which type of bonding is responsible for the attraction of water molecules to one another?

(A) covalent
(B) ionic
(C) metallic
(D) van der Waals

SOLUTIONS

1. Several divisions of materials can occur during classification. One such division is into metals, polymers (that is, plastics), and ceramics. Metal has the highest thermal and electrical conductivity of the material types.

The answer is (B).

2. Only bromine (Br) and mercury (Hg) are liquid under standard atmospheric conditions.

The answer is (C).

3. The approximate values for electrical conductivity at 20°C for the material options listed is shown in the table.

metal	S/m
aluminum	38×10^6
copper	58×10^6
gold	41×10^6
silver	63×10^6

The answer is (D).

4. Ceramic materials can be either crystalline or amorphous; exhibit poor toughness, that is, they tend to be brittle; and possess low electrical conductivity—hence their use as insulators.

The answer is (C).

5. Ionic, covalent, and metallic bonds are the three primary bonding methods. Bonds between molecules of the same substance are called van der Waals. There are three types of van der Waals forces; dipole to dipole (between polar molecules); dispersion (between non-polar molecules; and the hydrogen bond (also a dipole to dipole bond but stronger).

Thus, the van der Waals forces or bonding is responsible for the attraction of water molecules.

The answer is (D).

66 National Electrical Code

PRACTICE PROBLEMS

1. An unknown wire gage must be determined during an electrical inspection. The wire diameter is measured as 0.2591 cm. What is the AWG standard designation?

(A) AWG 4/0
(B) AWG 4
(C) AWG 8
(D) AWG 10

2. A given circuit is meant to carry a continuous lighting load of 16 A. In addition, four loads designed for permanent display stands are fastened in place and require 2 A each when operating. What is the rating of the overcurrent protective device (OCPD) on the branch circuit?

(A) 20 A
(B) 23 A
(C) 28 A
(D) 30 A

3. If standard three-conductor cable is used for the branch circuit in Prob. 66.2, what is the minimum conductor size?

(A) AWG 8
(B) AWG 10
(C) AWG 12
(D) AWG 14

4. A dwelling two-wire resistive heating circuit is designed for 240 V and 30 A. AWG 8 copper wire is used with a resistance of 0.809 Ω/1000 ft. What is the voltage drop for a one-way circuit length of 75 ft?

(A) 1.0%
(B) 1.5%
(C) 3.6%
(D) 5.0%

5. A three-wire feeder, AWG 1/0, carries 100 A of continuous load from a 480 V, 3ϕ (three-phase), three-wire source. The THHN copper conductors operate at 75°C and are enclosed within aluminum conduit. The expected power factor is 0.8. The total circuit length is 150 m. What is the voltage drop?

(A) 4.0 V
(B) 7.0 V
(C) 11 V
(D) 14 V

6. A 120 V dwelling branch circuit supplies four outlets, one of which has four receptacles. What is the total volt-ampere load for the circuit shown?

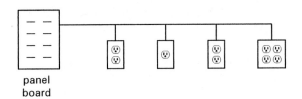

panel
board

(A) 180 VA
(B) 360 VA
(C) 720 VA
(D) 900 VA

7. A three-phase, four-wire feeder with a full-sized neutral in a separate raceway carries 14 A continuous and 40 A noncontinuous loads. The feeder uses an overcurrent device with a terminal rating of 75°C. What is the minimum copper conductor size?

(A) AWG 3
(B) AWG 4
(C) AWG 6
(D) AWG 8

8. All of the following are properties of fuses rather than circuit breakers with one exception. Which item is the property of circuit breakers?

(A) adjustable time-current characteristics
(B) low initial costs
(C) longer downtime after operation
(D) simplicity

Codes and
Standards

9. Installed equipment within a dwelling is attached to an overcurrent device (breaker) rated at 15 A. What is the required size of the equipment's grounding copper conductor?

(A) AWG 10
(B) AWG 12
(C) AWG 14
(D) AWG 18

10. A three-phase, four-wire feeder with termination provisions designed to carry less than 100 A uses THWN AWG 8 copper wire in an ambient operating environment of 50°C. From NEC Table 310.15(b)(2)(a), the derating factor for four conductors in a single raceway is 0.8. What is the allowable ampacity?

(A) 19 A
(B) 24 A
(C) 30 A
(D) 40 A

SOLUTIONS

1. The wire diameter is

$$d = (0.2591 \text{ cm}) \left(\frac{1 \text{ in}}{2.54 \text{ cm}} \right) = 0.1020 \text{ in}$$

Using this to enter a table of wire gages, such as the one in App. 66.A, indicates a 10 gage wire. That is, AWG 10.

The answer is (D).

2. From Article 210.20(A) of the National Electrical Code (NEC), the overcurrent protective device (OCPD) must be rated at 100% of the noncontinuous load plus 125% of the continuous load.

$$\text{OCPD} = (1.00)(2 \text{ A} + 2 \text{ A} + 2 \text{ A} + 2 \text{ A}) + (1.25)(16 \text{ A})$$
$$= 28 \text{ A minimum}$$

A standard fixed-trip circuit breaker or a fuse rated at 30 A can be used (see Sec. 240.6).

The answer is (D).

3. Conductor size is based on 100% of the noncontinuous load plus 125% of the continuous load [Sec. 210.19(A)(1)]. From Prob. 66.2, the total load is 28 A. The ampacity cannot be less than the rating of the overcurrent device, 30 A in this case [Sec. 310.15(B)FPN(2)]. The ampacity of the conductor for a circuit of less than 100 A should be based on the 60°C column of NEC Table 310.16 (App. 66.B) [Sec. 110.14(C)(1)].

Using App. 66.B, an AWG 10 conductor is chosen. (Checking Article 240.3 of the Code, as is called for by the asterisk in App. 66.B, indicates that 10 gage wire cannot have overcurrent protection greater than 30 A. All requirements are met.) Note that any derating of the conductor ampacity requires the use of smaller gage (larger diameter) conductor. See the conductor factors at the bottom of App. 66.B.

The answer is (B).

4. A two-wire circuit is either single phase or DC. The heating is resistive, so pf = 1 if the circuit is AC, which is the most likely scenario. Regardless, the voltage drop is given by Ohm's law as

$$V = IR$$
$$= IR_l l$$

The voltage drop occurs on both wires, so a factor of two must be used.

$$V = IR_l 2l$$

Substituting gives

$$V = 2IR_l l$$

$$= (2)(30 \text{ A}) \left(\frac{0.809 \ \Omega}{1000 \text{ ft}} \right) (75 \text{ ft})$$

$$= 3.64 \text{ V}$$

As a percentage of the 240 V nominal voltage, the voltage drop is

$$\left(\frac{3.64 \text{ V}}{240 \text{ V}} \right) \times 100\% = \boxed{1.516\% \quad (1.5\%)}$$

The answer is (B).

The conductor ampacities can be found in App. 66.B (NEC Table 310.16). Conductor properties, such as resistance (Ω) per 1000 ft, are found in App. 66.C (NEC Ch. 9, Table 8).

5. Note that although the continuous load is 100 A, and the voltage drop is calculated using that value, the conductors and the overcurrent protector must be able to carry 125 A, that is, 125% of the continuous load [Secs. 215-2(a) and 215-3].

The AC voltage drop is

$$V_{\text{drop}} = IZ$$

The effective impedance, corrected for the power factor, is calculated using App. 66.D (NEC Ch. 9, Table 9). For a 1/0 gage feeder, the resistance to neutral per 1000 ft in aluminum conduit is 0.13 Ω.

The inductive reactance, X_L, to neutral per 1000 ft is 0.044 Ω. The columns titled "alternating-current resistance for uncoated copper wires" and "X_L (reactance) for all wires" are used. Note that the uncoated copper wire column is used, although the THHN conductors are insulated. Also, the capacitive reactance is ignored. Thus, the voltage drop calculated will be an approximate value.

Because the power factor is not 0.85, the effective impedance cannot be taken directly from the table. The corrected impedance, Z_c, is

$$Z_c = R_l(\text{pf}) + X_{L,l} \sin(\arccos \text{pf})$$

$$= \left(0.13 \ \frac{\Omega}{1000 \text{ ft}} \right) (0.8)$$

$$+ \left(0.044 \ \frac{\Omega}{1000 \text{ ft}} \right) \sin 36.86°$$

$$= 0.1304 \ \Omega/1000 \text{ ft}$$

Because this is an "ohms-to-neutral" value, the line-to-neutral, or phase, voltage drop is

$$V_{\text{drop (line-to-neutral)}} = V_\phi = IZ_c$$

$$= (100 \text{ A}) \left(0.1304 \ \frac{\Omega}{1000 \text{ ft}} \right)$$

$$= 13.04 \text{ V}/1000 \text{ ft}$$

Adjusting for the total circuit length gives

$$V_\phi = \left(\frac{13.04 \text{ V}}{1000 \text{ ft}} \right) (150 \text{ m}) \left(\frac{1 \text{ ft}}{0.3048 \text{ m}} \right)$$

$$= 6.417 \text{ V}$$

Voltage drops are given in terms of line quantities for a three-phase, three-wire system.

$$V_{\text{line}} = V_\phi \sqrt{3}$$

$$= (6.417 \text{ V}) \sqrt{3}$$

$$= \boxed{11.1 \text{ V} \quad (11 \text{ V})}$$

This is approximately 2.5% of the 480 V source and is within NEC recommendations.

The answer is (C).

6. The situation is as shown.

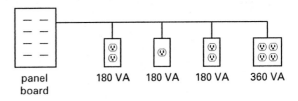

On any outlet with less than four receptacles, the VA load is 180 VA [Sec. 220.14(I)]. For four or more receptacles, the load is

$$VA = (\text{numbers of receptacles}) \left(90 \ \frac{VA}{\text{receptacle}} \right)$$

$$= (4)(90 \text{ VA})$$

$$= 360 \text{ VA}$$

$$VA_{\text{total}} = 180 \text{ VA} + 180 \text{ VA} + 180 \text{ VA} + 360 \text{ VA}$$

$$= \boxed{900 \text{ VA}}$$

The answer is (D).

7. Feeder conductor size, before derating, is based on 100% of the noncontinuous load and 125% of the continuous load [Art. 215.2(A)(1)].

$$A_{load} = (1.00)(40 \text{ A}) + (1.25)(14 \text{ A})$$
$$= 40 \text{ A} + 17.5 \text{ A}$$
$$= 57.5 \text{ A} \quad (58 \text{ A})$$

The total of 58 A is used because an ampacity of 0.5 A or greater is rounded up [Sec. 220.5(B)]. Using App. 66.B (NEC Table 310.16), select AWG 6 from the 75°C column. Always consider additional requirements or limits on the ratings. In this case, for circuits less than 100 A, the 60°C column is required [Sec. 110.14(C)(1)]. (Because of this article, the column primarily used in dwelling calculations is the one titled "60°C (140°F)".) Therefore, the AWG 6 conductor cannot be used as it is rated for 55 A at 60°C. Instead, AWG 4 with a 70 A rating is used.

The answer is (B).

8. The electrical characteristics of fuses are set by the material used, and no external adjustments can occur. The time-current characteristics cannot change.

The answer is (A).

9. Equipment grounding conductor sizes are given in NEC Table 250.122. For a 15 A overcurrent device, the grounding conductor must be AWG 14 copper.

The answer is (C).

10. Because the circuit carries less than 100 A and the termination provisions are designed commensurate with this condition, the 60°C column of App. 66.B must be used [see Sec. 110.14(C)(1)]. 40 A is the limiting allowable ampacity. The operating environment is above 30°C for which the values are calculated. THWN wire is rated for 75°C and 50 A. Using the correction factor table in App. 66.B, 0.75 is found in the 46–50°C row. The allowable ampacity in a given environment is determined by multiplying the rated ampacity in the table by all the correction factors—in this case, one for temperature and one for more than three conductors in a raceway.

$$A_{total} = A_{rated} F_{temp} F_{cond}$$
$$= (50 \text{ A})(0.75)(0.8)$$
$$= \boxed{30 \text{ A}}$$

Because 30 A is less than the 40 A maximum limit from the 60°C column, 30 A is the correct value.

The answer is (C).

67 National Electrical Safety Code

PRACTICE PROBLEMS

1. Street and area lights under the control of utilities are covered under the rules of the
- (A) ANSI
- (B) IEEE
- (C) NEC
- (D) NESC

2. What is the approximate required minimum clearance distance, R, if the phase voltage of the live parts is 13,800 V?

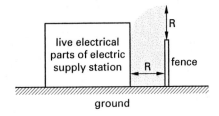

- (A) 1 m
- (B) 4 m
- (C) 10 m
- (D) 30 m

3. The New England states are considered to be in the _____ weather zone, according to Rule 250, and thus must meet the clearance requirements of zone _____ of Rule 230.
- (A) medium, 1
- (B) medium, 2
- (C) heavy, 1
- (D) heavy, 2

4. What approximate vertical clearance is required for an open supply conductor at 13.8 kV over a body of water (> 8 km² surface area) suitable for sail boating?
- (A) 4 m
- (B) 8 m
- (C) 12 m
- (D) 30 m

5. The wind load in newtons, L_{SI}, on an overhead line structure is calculated using the following formula.

$$L_{SI} = 0.613 V_{m/s}^2 k_z G_{RF} I C_f A_{m^2}$$

What does the k_z term represent?
- (A) force coefficient
- (B) gust response factor
- (C) velocity pressure coefficient
- (D) wind speed

SOLUTIONS

1. The NESC (National Electrical Safety Code contains the governing requirements for lights under the exclusive control of utilities (or their authorized contractors).

The answer is (D).

2. According to NESC Rule 110 and Table 110-1, for a voltage of 13,800 V, dimension R must be at least 3.1 m (4 m) or 10.1 ft (10 ft).

The answer is (B).

3. The New England states are in the heavy weather zone. The heavy, medium, and light weather conditions of Rule 250 correspond to clearance zones 1, 2, and 3, respectively, of Rule 230.

The answer is (C).

4. According to Rule 232, Table 232-1, of the NESC (row 7.d, column 5), the necessary clearance is 12.3 m (12 m) based on the high water mark.

The answer is (C).

5. The formula shown, along with the definitions of each term, is given in NESC Rule 250.C. The term k_z is the velocity pressure exposure coefficient (see Rule 250.C.1, Table 250-2).

The answer is (C).

68 Engineering Economic Analysis

PRACTICE PROBLEMS

1. At 6% effective annual interest, how much will be accumulated if $1000 is invested for ten years?

2. At 6% effective annual interest, what is the present worth of $2000 that becomes available in four years?

3. At 6% effective annual interest, how much should be invested to accumulate $2000 in 20 years?

4. At 6% effective annual interest, what year-end annual amount deposited over seven years is equivalent to $500 invested now?

5. At 6% effective annual interest, what will be the accumulated amount at the end of ten years if $50 is invested at the end of each year for ten years?

6. At 6% effective annual interest, how much should be deposited at the start of each year for ten years (a total of ten deposits) in order to empty the fund by drawing out $200 at the end of each year for ten years (a total of ten withdrawals)?

7. At 6% effective annual interest, how much should be deposited at the start of each year for five years to accumulate $2000 on the date of the last deposit?

8. At 6% effective annual interest, how much will be accumulated in ten years if three payments of $100 are deposited every other year for four years, with the first payment occurring at $t = 0$?

9. $500 is compounded monthly at a 6% nominal annual interest rate. How much will have accumulated in five years?

10. What is the effective annual rate of return on an $80 investment that pays back $120 in seven years?

11. A new machine will cost $17,000 and will have a resale value of $14,000 after five years. Special tooling will cost $5000. The tooling will have a resale value of $2500 after five years. Maintenance will be $2000 per year. The effective annual interest rate is 6%. What will be the average annual cost of ownership during the next five years?

12. An old covered wooden bridge can be strengthened at a cost of $9000, or it can be replaced for $40,000. The present salvage value of the old bridge is $13,000. It is estimated that the reinforced bridge will last for 20 years, will have an annual cost of $500, and will have a salvage value of $10,000 at the end of 20 years. The estimated salvage value of the new bridge after 25 years is $15,000. Maintenance for the new bridge would cost $100 annually. The effective annual interest rate is 8%. Which is the better alternative?

13. A firm expects to receive $32,000 each year for 15 years from sales of a product. An initial investment of $150,000 will be required to manufacture the product. Expenses will run $7530 per year. The salvage value is zero, and straight line depreciation is used. The income tax rate is 48%. What is the after-tax rate of return?

14. A public works project has initial costs of $1,000,000, benefits of $1,500,000, and disbenefits of $300,000. (a) What is the benefit/cost ratio? (b) What is the excess of benefits over costs?

15. A speculator in land pays $14,000 for property that he expects to hold for ten years. $1000 is spent in renovation, and a monthly rent of $75 is collected from the tenants. (Use the year-end convention.) Taxes are $150 per year, and maintenance costs are $250 per year. What must be the sale price in ten years to realize a 10% rate of return?

16. What is the effective annual interest rate for a payment plan of 30 equal payments of $89.30 per month when a lump sum payment of $2000 would have been an outright purchase?

17. A depreciable item is purchased for $500,000. The salvage value at the end of 25 years is estimated at $100,000. What is the depreciation in each of the first three years using the (a) straight line, (b) sum-of-the-years' digits, and (c) double declining balance methods?

18. Equipment that is purchased for $12,000 now is expected to be sold after ten years for $2000. The estimated maintenance is $1000 for the first year, but it is expected to increase $200 each year thereafter. The effective annual interest rate is 10%. What are the (a) present worth and (b) annual cost?

19. A new grain combine with a 20-year life can remove 7 lbm of rocks from its harvest per hour. Any rocks left in its output hopper will cause $25,000 damage in subsequent processes. Several investments are available to increase the rock-removal capacity, as listed in the table. The effective annual interest rate is 10%. What should be done?

rock removal rate	probability of exceeding rock removal rate	required investment to achieve removal rate ($)
7	0.15	0
8	0.10	15,000
9	0.07	20,000
10	0.03	30,000

20. A mechanism that costs $10,000 has operating costs and salvage values as given. An effective annual interest rate of 20% is to be used.

year	operating cost ($)	salvage value ($)
1	2000	8000
2	3000	7000
3	4000	6000
4	5000	5000
5	6000	4000

(a) What is the economic life of the mechanism? (b) Assuming that the mechanism has been owned and operated for four years already, what is the cost of owning and operating the mechanism for one more year?

21. A salesperson intends to purchase a car for $50,000 for personal use, driving 15,000 miles per year. Insurance for personal use costs $2000 per year, and maintenance costs $1500 per year. The car gets 15 miles per gallon, and gasoline costs $1.50 per gallon. The resale value after five years will be $10,000. The salesperson's employer has asked that the car be used for business driving of 50,000 miles per year and has offered a reimbursement of $0.30 per mile. Using the car for business would increase the insurance cost to $3000 per year and maintenance to $2000 per year. The salvage value after five years would be reduced to $5000. If the employer purchased a car for the salesperson to use, the initial cost would be the same, but insurance, maintenance, and salvage would be $2500, $2000, and $8000, respectively. The salesperson's effective annual interest rate is 10%. (a) Is the reimbursement offer adequate? (b) With a reimbursement of $0.30 per mile, how many miles must the car be driven per year to justify the employer buying the car for the salesperson to use?

22. Alternatives A and B are being evaluated. The effective annual interest rate is 10%. Which alternative is economically superior?

	alternative A	alternative B
initial cost ($)	80,000	35,000
life (yr)	20	10
salvage value ($)	7000	0
annual costs ($)		
years 1–5	1000	3000
years 6–10	1500	4000
years 11–20	2000	0
additional cost ($)		
year 10	5000	0

23. A car is needed for three years. Plans A and B for acquiring the car are being evaluated. An effective annual interest rate of 10% is to be used. Which plan is economically superior?

plan A: Lease the car for $0.25/mile (all inclusive).

plan B: Purchase the car for $30,000, keep the car for three years, and sell the car after three years for $7200. Pay $0.14 per mile for oil and gas. Pay other costs of $500 per year.

24. Two methods are being considered to meet strict air pollution control requirements over the next ten years. Method A uses equipment with a life of ten years. Method B uses equipment with a life of five years that will be replaced with new equipment with an additional life of five years. Capacities of the two methods are different, but operating costs do not depend on the throughput. Operation is 24 hours per day, 365 days per year. The effective annual interest rate for this evaluation is 7%.

	method A	method B	
	years 1–10	years 1–5	years 6–10
installation cost ($)	13,000	6000	7000
equipment cost ($)	10,000	2000	2200
operating cost ($/hr)			
per hour	10.50	8.00	8.00
salvage value ($)	5000	2000	2000
capacity (tons/yr)	50	20	20
life (yr)	10	5	5

(a) What is the uniform annual cost per ton for each method? (b) Over what range of throughput (in tons/yr) does each method have the minimum cost?

25. A transit district has asked for assistance in determining the proper fare for its bus system. An effective annual interest rate of 7% is to be used. The following additional information was compiled for the study.

cost ($/bus)	60,000
bus life (yr)	20
salvage value ($)	10,000
distance driven (mi/yr)	37,440
number of passengers per year	80,000
operating cost ($/mi)	1 (increasing $0.10 per mile each year after the first)

(a) If the fare is to remain constant for the next 20 years, what is the break-even fare per passenger? (b) If the transit district decides to set the per-passenger fare at $0.35 for the first year, by what amount should the per-passenger fare go up each year thereafter such that the district can break even in 20 years? (c) If the transit district decides to set the per-passenger fare at $0.35 for the first year and the per-passenger fare goes up $0.05 each year thereafter, what additional governmental subsidy (per passenger) is needed for the district to break even in 20 years?

26. Make a recommendation to a client to accept one of the following alternatives. Use the present worth comparison method. (Initial costs are the same.)

alternative A: a 25-year annuity paying $4800 at the end of each year, where the interest rate is a nominal 12% per annum

alternative B: a 25-year annuity paying $1200 every quarter at 12% nominal annual interest

27. A firm has two alternatives for improvement of its existing production line. The data are as follows.

	alternative A	alternative B
initial installation cost ($)	1500	2500
annual operating cost ($)	800	650
service life (yr)	5	8
salvage value ($)	0	0

Determine the best alternative using an interest rate of 15%.

28. Two mutually exclusive alternatives requiring different investments are being considered. The lives of both alternatives are estimated at 20 years with no salvage values. The minimum rate of return that is considered acceptable is 4%. Which alternative is better?

	alternative A	alternative B
investment required ($)	70,000	40,000
net income ($/yr)	5620	4075
rate of return on total investment (%)	5	8

29. Compare the costs of two plant renovation schemes, alternatives A and B. Assume equal lives of 25 years, no salvage values, and interest at 25%. Make the comparison on the basis of (a) present worth, (b) capitalized cost, and (c) annual cost.

	alternative A	alternative B
initial cost ($)	20,000	25,000
annual expenditure ($/yr)	3000	2500

30. With interest at 8%, obtain the solutions to the following to the nearest dollar. (a) A machine costs $18,000 and has a salvage value of $2000. It has a useful life of eight years. What is its book value at the end of five years using straight line depreciation? (b) Using data from part (a), find the depreciation in the first three years using the sinking fund method. (c) Repeat part (a) using double declining balance depreciation to find the first five years' depreciation.

31. A chemical pump motor unit is purchased for $14,000. The estimated life is eight years, after which it will be sold for $1800. Find the depreciation in the first two years by the sum-of-the-years' digits method. Calculate the after-tax depreciation recovery using 15% interest with 52% income tax.

32. A soda ash plant has the water effluent from processing equipment treated in a large settling basin. The settling basin eventually discharges into a river that runs alongside the basin. Recently enacted environmental regulations require all rainfall on the plant to be diverted and treated in the settling basin. A heavy rainfall will cause the entire basin to overflow. An uncontrolled overflow will cause environmental damage and heavy fines. The construction of additional height on the existing basic walls is under consideration.

Data on the costs of construction and expected costs for environmental cleanup and fines are shown. Data on 50 typical winters have been collected. The soda ash plant management considers 12% to be their minimum rate of return, and they predict that after 15 years the plant will be closed. The company wants to select the alternative that minimizes its total expected costs. How high should the walls be built?

additional basin height (ft)	number of winters with basin overflow	expense for environmental cleanup ($/yr)	construction cost ($)
0	24	550,000	0
5	14	600,000	600,000
10	8	650,000	710,000
15	3	700,000	900,000
20	1	800,000	1,000,000
	50		

33. A wood processing plant installed a waste gas scrubber at a cost of $30,000 to remove pollutants from the exhaust discharged into the atmosphere. The scrubber has no salvage value and will cost $18,700 to operate next year, with operating costs expected to increase at the rate of $1200 per year thereafter. Money can be borrowed at 12%. When should the company consider replacing the scrubber?

34. Two alternative piping schemes are being considered by a water treatment facility. On the basis of a ten-year life and an interest rate of 12%, determine the number of hours of operation for which the two installations will be equivalent.

	alternative A	alternative B
pipe diameter (in)	4	6
head loss for required flow (ft)	48	26
size motor required (hp)	20	7
energy cost per hour of operation ($)	0.30	0.10
cost of motor installed ($)	3600	2800
cost of pipes and fittings ($)	3050	5010
salvage value at end of ten years ($)	200	280

35. An 88% learning curve is used with an item whose first production time was six weeks. (a) How long will it take to produce the fourth item? (b) How long will it take to produce the sixth through fourteenth items?

36. A company is considering two alternatives, only one of which can be selected.

alternative	initial investment ($)	salvage value ($)	annual net profit ($/yr)	life (yr)
A	120,000	15,000	57,000	5
B	170,000	20,000	67,000	5

The net profit is after operating and maintenance costs but before taxes. The company pays 45% of its year-end profit as income taxes. Use straight line depreciation. Do not use investment tax credit. Find the best alternative if the company's minimum attractive rate of return is 15%.

37. A company is considering the purchase of equipment to expand its capacity. The equipment cost is $300,000. The equipment is needed for five years, after which it will be sold for $50,000. The company's before-tax cash flow will be improved $90,000 annually by the purchase of the asset.

The corporate tax rate is 48%, and straight line depreciation will be used. The company will take an investment tax credit of 6.67%. What is the after-tax rate of return associated with this equipment purchase?

38. A 120-room hotel is purchased for $2,500,000. A 25-year loan is available for 12%. A study was conducted to determine the various occupancy rates.

occupancy (% full)	probability
65	0.40
70	0.30
75	0.20
80	0.10

The operating costs of the hotel are as follows.

taxes and insurance ($/yr)	20,000
maintenance ($/yr)	50,000
operating ($/yr)	200,000

The life of the hotel is projected to be 25 years when operating 365 days per year. The salvage value after 25 years is $500,000.

Neglect tax credit and income taxes. Determine the average rate that should be charged per room per night to return 15% of the initial cost each year.

39. A company is insured for $3,500,000 against fire. The insurance rate is $0.69/1000. The insurance company will decrease the rate to $0.47/1000 if fire sprinklers are installed. The initial cost of the sprinklers is $7500. Annual costs are $200; additional taxes are $100 annually. The system life is 25 years. What is the rate of return on this investment?

40. Heat losses through the walls in an existing building cost a company $1,300,000 per year. This amount is considered excessive, and two other options are being evaluated. Neither of the options will increase the life of the existing building beyond the current expected life of six years, and neither of the alternatives will produce a salvage value.

alternative A: Do nothing. Continue with current losses.

alternative B: Spend $2,000,000 immediately to upgrade the building and reduce the loss by 80%. This alternative will require annual maintenance costing $150,000.

alternative C: Spend $1,200,000 immediately. Repeat the $1,200,000 expenditure three years from now. Heat loss the first year will be reduced 80%. Due to deterioration, the reduction will be 55% and 20% in the second and third years. (The pattern is repeated starting after the second expenditure.) There are no maintenance costs.

All energy and maintenance costs are considered expenses for tax purposes. The company's tax rate is 48%, and straight line depreciation is used. 15% is regarded as the effective annual interest rate. Evaluate each alternative on an after-tax basis, and recommend the best alternative.

41. Determine if a seven-year-old machine should be replaced. Give a full explanation. Use a before-tax interest rate of 15%.

The existing machine is presumed to have a ten-year life. It has been depreciated on a straight line basis from its original value of $1,250,000 to a current book value of $620,000. Its ultimate salvage value was assumed to be $350,000 for purposes of depreciation. Its present salvage value is estimated at $400,000, and this is not expected to change over the next three years. The current operating costs are not expected to change from $200,000 per year.

A new machine costs $800,000, with operating costs of $40,000 the first year, and increasing by $30,000 each year thereafter. The new machine has an expected life of ten years. The salvage value depends on the year the new machine is retired.

year retired	salvage value ($)
1	600,000
2	500,000
3	450,000
4	400,000
5	350,000
6	300,000
7	250,000
8	200,000
9	150,000
10	100,000

SOLUTIONS

1.

$i = 6\%/\text{yr}$

By the formula from Table 68.1,

$$F = P(1+i)^n = (\$1000)(1+0.06)^{10}$$

$$= \boxed{\$1790.85}$$

By the factor converting P to F, $(F/P, i, n) = 1.7908$ for an interest rate of $i = 6\%/\text{yr}$ and a time of $n = 10$ yr.

$$F = P(F/P, 6\%, 10) = (\$1000)(1.7908)$$

$$= \boxed{\$1790.80}$$

2.

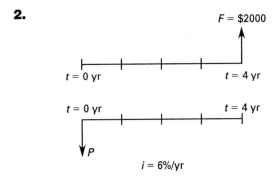

$i = 6\%/\text{yr}$

By the formula from Table 68.1,

$$P = \frac{F}{(1+i)^n} = \frac{\$2000}{(1+0.06)^4}$$

$$= \boxed{\$1584.19}$$

From the factor converting F to P, $(P/F, i, n) = 0.7921$ for $i = 6\%/\text{yr}$ and $n = 4$ yr.

$$P = F(P/F, 6\%, 4) = (\$2000)(0.7921)$$

$$= \boxed{\$1584.20}$$

3.

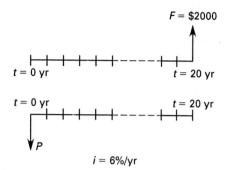

$i = 6\%/\text{yr}$

By the formula from Table 68.1,

$$P = \frac{F}{(1+i)^n} = \frac{\$2000}{(1+0.06)^{20}}$$

$$= \boxed{\$623.61}$$

From the factor converting F to P, $(P/F, i, n) = 0.3118$ for $i = 6\%/\text{yr}$ and $n = 20$ yr.

$$P = F(P/F, 6\%, 20) = (\$2000)(0.3118)$$

$$= \boxed{\$623.60}$$

4.

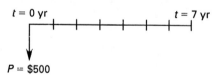

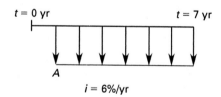

$i = 6\%/\text{yr}$

By the formula from Table 68.1,

$$A = P\left(\frac{i(1+i)^n}{(1+i)^n - 1}\right) = (\$500)\left(\frac{(0.06)(1+0.06)^7}{(1+0.06)^7 - 1}\right)$$

$$= \boxed{\$89.57}$$

By the factor converting P to A, $(A/P, i, n) = 0.17914$ for $i = 6\%/\text{yr}$ and $n = 7$ yr.

$$A = P(A/P, 6\%, 7) = (\$500)(0.17914)$$

$$= \boxed{\$89.57}$$

5.

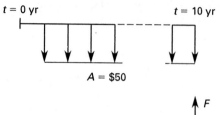

$A = \$50$

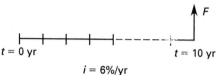

$i = 6\%/\text{yr}$

By the formula from Table 68.1,

$$F = A\left(\frac{(1+i)^n - 1}{i}\right) = (\$50)\left(\frac{(1+0.06)^{10} - 1}{0.06}\right)$$

$$= \boxed{\$659.04}$$

By the factor converting A to F, $(F/A, i, n) = 13.181$ for $i = 6\%/\text{yr}$ and $n = 10$ yr.

$$F = A(F/A, 6\%, 10) = (\$50)(13.181)$$

$$= \boxed{\$659.05}$$

6.

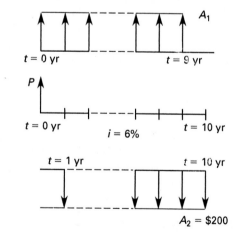

By the formula from Table 68.1, for each cash flow diagram,

$$P = (A_1 + A_1)\left(\frac{(1+0.06)^9 - 1}{(0.06)(1+0.06)^9}\right)$$

$$= A_2\left(\frac{(1+0.06)^{10} - 1}{(0.06)(1+0.06)^{10}}\right)$$

Therefore, for $A_2 = \$200$,

$$A_1 + A_1\left(\frac{(1+0.06)^9 - 1}{(0.06)(1+0.06)^9}\right)$$

$$= (\$200)\left(\frac{(1+0.06)^{10} - 1}{(0.06)(1+0.06)^{10}}\right)$$

$$7.80A_1 = \$1472.02$$

$$\boxed{A_1 = \$188.72}$$

By the factor converting A to P,

$$(P/A, 6\%, 9) = 6.8017$$

$$(P/A, 6\%, 10) = 7.3601$$

$$A_1 + A_1(6.8017) = (\$200)(7.3601)$$

$$7.8017A_1 = \$1472.02$$

$$A_1 = \frac{\$1472.02}{7.8017}$$

$$\boxed{= \$188.68}$$

7.

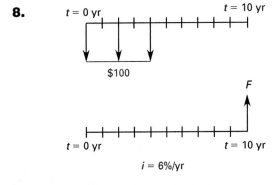

$t = 0$ yr $\qquad$ $t = 4$ yr

A

$F = \$2000$

$t = 0$ yr $\qquad$ $t = 4$ yr

$i = 6\%$/yr

By the formula from Table 68.1,

$$F = A\left(\frac{(1+i)^n - 1}{i}\right)$$

Because the deposits start at the start of each year, $n = 4 + 1$, for a total of five deposits.

$$F = A\left(\frac{(1+i)^{n+1} - 1}{i}\right) = \$2000$$

$$= A\left(\frac{(1+0.06)^5 - 1}{0.06}\right)$$

$$\$2000 = 5.6371A$$

$$A = \frac{\$2000}{5.6371}$$

$$\boxed{= \$354.79}$$

By the factor converting P and A to F,

$$F = A((F/P, 6\%, 4) + (F/A, 6\%, 4))$$

$$\$2000 = A(1.2625 + 4.3746)$$

$$\boxed{A = \$354.79}$$

8.

$t = 0$ yr $\qquad$ $t = 10$ yr

$\$100$

F

$t = 0$ yr $\qquad$ $t = 10$ yr

$i = 6\%$/yr

By the formula from Table 68.1, $F = P(1+i)^n$. If each deposit is considered as P, each will accumulate interest for periods of ten, eight, and six years.

Therefore,

$$F = (\$100)(1+0.06)^{10} + (\$100)(1+0.06)^8$$

$$+ (\$100)(1+0.06)^6$$

$$= (\$100)(1.7908 + 1.5938 + 1.4185)$$

$$\boxed{= \$480.31}$$

By the factor converting P to F,

$$(F/P, i, n) = 1.7908 \text{ for } i = 6\% \text{ and } n = 10$$

$$= 1.5938 \text{ for } i = 6\% \text{ and } n = 8$$

$$= 1.4185 \text{ for } i = 6\% \text{ and } n = 6$$

By summation,

$$F = (\$100)(1.7908 + 1.5938 + 1.4185)$$

$$\boxed{= \$480.31}$$

9.

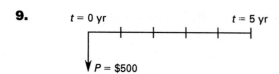

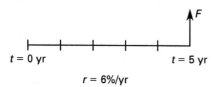

Because the deposit is compounded monthly, the effective interest rate should be calculated as shown by Eq. 68.54.

$$i = \left(1 + \frac{r}{k}\right)^k - 1 = \left(1 + \frac{0.06}{12}\right)^{12} - 1$$

$$= 0.061677 \quad (6.1677\%)$$

By the formula from Table 68.1,

$$F = P(1 + i)^n = (\$500)(1 + 0.061677)^5$$

$$= \boxed{\$674.42}$$

To use a table of factors, interpolation is required.

$i\%$	factor F/P
6	1.3382
6.1677	desired
7	1.4026

$$i = \left(\frac{6.1677 - 6}{7 - 6}\right)(1.4026 - 1.3382)$$

$$= 0.0108$$

Therefore,

$$F/P = 1.3382 + 0.0108 = 1.3490$$

$$F = P(F/P, 6.1677\%, 5) = (\$500)(1.3490)$$

$$= \boxed{\$674.50}$$

10.

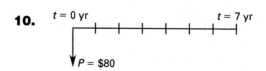

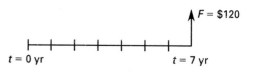

By the formula from Table 68.1,

$$F = P(1 + i)^n$$

Therefore,

$$(1 + i)^n = F/P$$

$$i = (F/P)^{1/n} - 1 = \left(\frac{\$120}{\$80}\right)^{1/7} - 1$$

$$= \boxed{0.059 \quad (6\%)}$$

By the factor coverting P to F,

$$F = P(F/P, i\%, 7)$$

$$(F/P, i\%, 7) = F/P = \frac{\$120}{\$80} = 1.5$$

By checking the interest tables,

$$(F/P, i\%, 7) = 1.4071 \text{ for } i = 5\%$$

$$= 1.5036 \text{ for } i = 6\%$$

$$= 1.6058 \text{ for } i = 7\%$$

Therefore, $i \approx \boxed{6\%}$.

11.

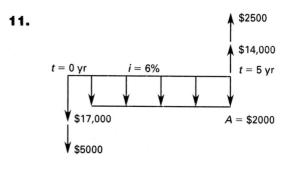

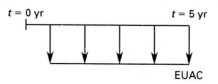

Annual cost of ownership, EUAC, can be obtained from the factors converting P to A and F to A.

$$P = \$17,000 + \$5000$$

$$= \$22,000$$

$$F = \$14,000 + \$2500$$

$$= \$16,500$$

$$EUAC = A + P(A/P, 6\%, 5) - F(A/F, 6\%, 5)$$

$$(A/P, 6\%, 5) = 0.23740$$

$$(A/F, 6\%, 5) = 0.17740$$

$$EUAC = \$2000 + (\$22,000)(0.23740)$$
$$- (\$16,500)(0.17740)$$

$$= \boxed{\$4295.70}$$

12.

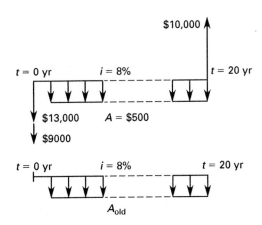

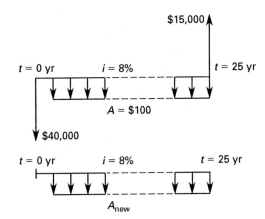

Consider the salvage value as a benefit lost (cost).

$$EUAC_{old} = \$500 + (\$22,000)(A/P, 8\%, 20)$$
$$- (\$10,000)(A/F, 8\%, 20)$$

$$(A/P, 8\%, 20) = 0.1019$$

$$(A/F, 8\%, 20) = 0.0219$$

$$EUAC_{old} = \$500 + (\$22,000)(0.1019)$$
$$- (\$10,000)(0.0219)$$

$$= \$2522.80$$

Similarly,

$$EUAC_{new} = \$100 + (\$40,000)(A/P, 8\%, 25)$$
$$- (\$15,000)(A/F, 8\%, 25)$$

$$(A/P, 8\%, 25) = 0.0937$$

$$(A/F, 8\%, 25) = 0.0137$$

$$EUAC_{new} = \$100 + (\$40,000)(0.0937)$$
$$- (\$15,000)(0.0137)$$

$$= \$3642.50$$

Therefore, the new bridge is going to be more costly.

$$\boxed{\text{The best alternative is to strengthen the old bridge.}}$$

13.

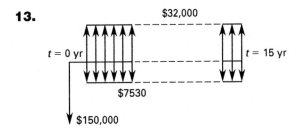

The annual depreciation is

$$D = \frac{C - S_n}{n} = \frac{\$150,000 - \$0}{15 \text{ yr}}$$

$$= \$10,000/\text{yr}$$

The taxable income is

$$\$32,000 - \$7530 - \$10,000 = \$14,470/\text{yr}$$

Taxes paid are

$$(\$14,470)(0.48) = \$6945.60/\text{yr}$$

The after-tax cash flow is

$$\$24,470 - \$6945.60 = \$17,524.40$$

The present worth of the alternate is zero when evaluated at its ROR.

$$0 = -\$150,000 + (\$17,524.40)(P/A, i\%, 15)$$

Therefore,

$$(P/A, i\%, 15) = \frac{\$150,000}{\$17,524.40} = 8.55949$$

In the tables, this factor matches $i = 8\%$.

$$\boxed{\text{ROR} = 8\%}$$

14. The conventional benefit/cost ratio is

$$B/C = \frac{B - D}{C}$$

(a) The benefit/cost ratio is

$$B/C = \frac{\$1,500,000 - \$300,000}{\$1,000,000}$$

$$= \boxed{1.2}$$

(b) The excess of benefits over cost is

$$\boxed{\$200,000.}$$

15.

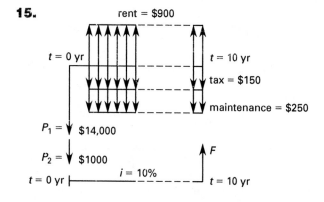

The annual rent is

$$(\$75)\left(12\ \frac{\text{mo}}{\text{yr}}\right) = \$900$$

$$P = P_1 + P_2 = \$15,000$$

$$A_1 = -\$900$$

$$A_2 = \$250 + \$150 = \$400$$

By the factors converting P to F and A to F,

$$F = (\$15,000)(F/P, 10\%, 10)$$
$$+ (\$400)(F/A, 10\%, 10)$$
$$- (\$900)(F/A, 10\%, 10)$$

$$(F/P, 10\%, 10) = 2.5937$$

$$(F/A, 10\%, 10) = 15.937$$

$$F = (\$15,000)(2.5937)$$
$$+ (\$400)(15.937)$$
$$- (\$900)(15.937)$$

$$= \boxed{\$30,937}$$

16.

$t = 0$ yr, $t = 30$ yr, $A = \$89.30$

$t = 0$ yr, $t = 30$ yr, $P = \$2000$

By the formula relating P to A,

$$P = A\left(\frac{(1+i)^n - 1}{i(1+i)^n}\right)$$

$$\frac{(1+i)^{30} - 1}{i(1+i)^{30}} = \frac{\$2000}{\$89.30} = 22.40$$

By trial and error,

i (%)	$(1+i)^{30}$	$\dfrac{(1+i)^{30} - 1}{i(1+i)^{30}}$
10	17.45	9.42
6	5.74	13.76
4	3.24	17.28
2	1.81	22.37

2% per month is close.

$$i = (1 + 0.02)^{12} - 1$$

$$= \boxed{0.2682 \quad (26.82\%)}$$

17. (a) Use the straight line method, Eq. 68.25.

$$D = \frac{C - S_n}{n}$$

Each year depreciation will remain the same.

$$D = \frac{\$500,000 - \$100,000}{25}$$

$$= \boxed{\$16,000}$$

(b) Sum-of-the-years' digits (SOYD) can be calculated as shown by Eq. 68.28.

$$D_j = \frac{(C - S_n)(n - j + 1)}{T}$$

Use Eq. 68.27.

$$T = \tfrac{1}{2}n(n + 1) = \left(\tfrac{1}{2}\right)(25)(25 + 1) = 325$$

$$D_1 = \frac{(\$500,000 - \$100,000)(25 - 1 + 1)}{325}$$

$$= \boxed{\$30,769}$$

$$D_2 = \frac{(\$500,000 - \$100,000)(25 - 2 + 1)}{325}$$

$$= \boxed{\$29,538}$$

$$D_3 = \frac{(\$500,000 - \$100,000)(25 - 3 + 1)}{325}$$

$$= \boxed{\$28,308}$$

(c) The double declining balance (DDB) method can be used. By Eq. 68.32,

$$D_j = dC(1 - d)^{j-1}$$

Use Eq. 68.31.

$$d = \frac{2}{n} = \frac{2}{25}$$

$$D_1 = \left(\tfrac{2}{25}\right)(\$500,000)\left(1 - \tfrac{2}{25}\right)^0 = \boxed{\$40,000}$$

$$D_2 = \left(\tfrac{2}{25}\right)(\$500,000)\left(1 - \tfrac{2}{25}\right)^1 = \boxed{\$36,800}$$

$$D_3 = \left(\tfrac{2}{25}\right)(\$500,000)\left(1 - \tfrac{2}{25}\right)^2 = \boxed{\$33,856}$$

18.

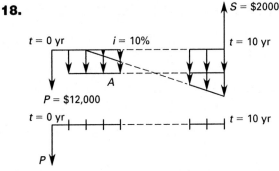

(a) $A = \$1000$ and $G = \$200$ for $t = n - 1 = 9$ yr.

$$F = S = \$2000$$

$$P = \$12,000 + A(P/A, 10\%, 10) + G(P/G, 10\%, 10)$$
$$\quad - F(P/F, 10\%, 10)$$

$$= \$12,000 + (\$1000)(6.1446) + (\$200)(22.8913)$$
$$\quad - (\$2000)(0.3855)$$

$$= \boxed{\$21,952}$$

(b) $A = (\$12,000)(A/P, 10\%, 10) + \$1000$
$$\quad + (\$200)(A/G, 10\%, 10)$$
$$\quad - (\$2000)(A/F, 10\%, 10)$$

$$= (\$12,000)(0.1627) + \$1000 + (\$200)(3.7255)$$
$$\quad - (\$2000)(0.0627)$$

$$= \boxed{\$3572.70}$$

19. An increase in rock removal capacity can be achieved by a 20-year loan (investment). Different cases available can be compared by equivalent uniform annual cost (EUAC).

$$EUAC = \text{annual loan cost}$$
$$\qquad + \text{expected annual damage}$$
$$\quad = \text{cost } (A/P, 10\%, 20)$$
$$\qquad + (\$25,000)(\text{probability})$$

$$(A/P, 10\%, 20) = 0.1175$$

A table can be prepared for different cases.

rock removal rate	cost ($)	annual loan cost ($)	expected annual damage ($)	EUAC ($)
7	0	0	3750	3750.00
8	15,000	1761.90	2500	4261.90
9	20,000	2349.20	1750	4099.20
10	30,000	3523.80	750	4273.80

It is cheapest to do nothing.

20. Calculate the cost of owning and operating for each year.

$$A_1 = (\$10,000)(A/P, 20\%, 1) + \$2000 - (\$8000)(A/F, 20\%, 1)$$

$$(A/P, 20\%, 1) = 1.2$$

$$(A/F, 20\%, 1) = 1.0$$

$$A_1 = (\$10,000)(1.2) + \$2000 - (\$8000)(1.0)$$

$$= \$6000$$

$$A_2 = (\$10,000)(A/P, 20\%, 2) + \$2000 + (\$1000)(A/G, 20\%, 2) - (\$7000)(A/F, 20\%, 2)$$

$$(A/P, 20\%, 2) = 0.6545$$

$$(A/G, 20\%, 2) = 0.4545$$

$$(A/F, 20\%, 2) = 0.4545$$

$$A_2 = (\$10,000)(0.6545) + \$2000 + (\$1000)(0.4545) - (\$7000)(0.4545)$$

$$= \$5818$$

$$A_3 = (\$10,000)(A/P, 20\%, 3) + \$2000 + (\$1000)(A/G, 20\%, 3) - (\$6000)(A/F, 20\%, 3)$$

$$(A/P, 20\%, 3) = 0.4747$$

$$(A/G, 20\%, 3) = 0.8791$$

$$(A/F, 20\%, 3) = 0.2747$$

$$A_3 = (\$10,000)(0.4747) + \$2000 + (\$1000)(0.8791) - (\$6000)(0.2747)$$

$$= \$5977.90$$

$$A_4 = (\$10,000)(A/P, 20\%, 4) + \$2000 + (\$1000)(A/G, 20\%, 4) - (\$5000)(A/F, 20\%, 4)$$

$$(A/P, 20\%, 4) = 0.3863$$

$$(A/G, 20\%, 4) = 1.2762$$

$$(A/F, 20\%, 4) = 0.1863$$

$$A_4 = (\$10,000)(0.3863) + \$2000 + (\$1000)(1.2762) - (\$5000)(0.1863)$$

$$= \$6207.70$$

$$A_5 = (\$10,000)(A/P, 20\%, 5) + \$2000 + (\$1000)(A/G, 20\%, 5) - (\$4000)(A/F, 20\%, 5)$$

$$(A/P, 20\%, 5) = 0.3344$$

$$(A/G, 20\%, 5) = 1.6405$$

$$(A/F, 20\%, 5) = 0.1344$$

$$A_5 = (\$10,000)(0.3344) + \$2000 + (\$1000)(1.6405) - (\$4000)(0.1344)$$

$$= \$6446.90$$

(a) Because the annual owning and operating cost is smallest after two years of operation, it is advantageous to sell the mechanism after the second year.

The economic life is two years.

(b) After four years of operation, the owning and operating cost of the mechanism for one more year is

$$A = \$6000 + (\$5000)(1 + i) - \$4000$$

$$i = 0.2 \ (20\%)$$

$$A = \$6000 + (\$5000)(1.2) - \$4000$$

$$= \boxed{\$8000}$$

21. To find out if the reimbursement is adequate, calculate the business-related expense.

Charge the company for business travel.

insurance: $3000 - \$2000 = \$1000$

maintenance: $2000 - \$1500 = \$500$

drop in salvage value: $10,000 - \$5000 = \$5000$

The annual portion of the drop in salvage value is

$$A = (\$5000)(A/F, 10\%, 5)$$

$$(A/F, 10\%, 5) = 0.1638$$

$$A = (\$5000)(0.1638) = \$819/\text{yr}$$

(a) The annual cost of gas is

$$\left(\frac{50,000 \text{ mi}}{15 \frac{\text{mi}}{\text{gal}}}\right)\left(1.50 \frac{\$}{\text{gal}}\right) = \$5000$$

$$\text{EUAC per mile} = \frac{\$1000 + \$500 + \$819 + \$5000}{50,000 \text{ mi}}$$

$$= \boxed{\$0.14638/\text{mi}}$$

Because the reimbursement per mile is \$0.30 and \$0.30 > \$0.14638, the reimbursement is adequate.

(b) Next, determine (with reimbursement) how many miles the car must be driven to break even.

If the car is driven M miles per year,

$$\left(0.30 \frac{\$}{\text{mi}}\right)M = (\$50,000)(A/P, 10\%, 5) + \$2500$$

$$+ \$2000 - (\$8000)(A/F, 10\%, 5)$$

$$+ \left(\frac{M}{15 \frac{\text{mi}}{\text{gal}}}\right)(\$1.50)$$

$$(A/P, 10\%, 5) = 0.2638$$

$$(A/F, 10\%, 5) = 0.1638$$

$$0.3M = (\$50,000)(0.2638) + \$2500 + \$2000$$

$$- (\$8000)(0.1638) + 0.1M$$

$$0.2M = \$16,379.60$$

$$M = \frac{\$16,379.60}{0.2 \frac{\$}{\text{mi}}}$$

$$= \boxed{81,898 \text{ mi}}$$

22.

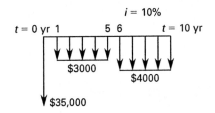

$$P_A = \$80,000 + (\$1000)(P/A, 10\%, 5)$$

$$+ (\$1500)(P/A, 10\%, 5)$$

$$\times (P/F, 10\%, 5)$$

$$+ (\$2000)(P/A, 10\%, 10)$$

$$\times (P/F, 10\%, 10)$$

$$+ (\$5000)(P/F, 10\%, 10)$$

$$- (\$7000)(P/F, 10\%, 20)$$

$$(P/A, 10\%, 5) = 3.7908$$

$$(P/F, 10\%, 5) = 0.6209$$

$$(P/A, 10\%, 10) = 6.1446$$

$$(P/F, 10\%, 10) = 0.3855$$

$$(P/F, 10\%, 20) = 0.1486$$

$$P_A = \$80,000 + (\$1000)(3.7908)$$

$$+ (\$1500)(3.7908)(0.6209)$$

$$+ (\$2000)(6.1446)(0.3855)$$

$$+ (\$5000)(0.3855)$$

$$- (\$7000)(0.1486)$$

$$= \$92,946.15$$

Because the lives are different, compare by EUAC.

$$\text{EUAC(A)} = (\$92,946.14)(A/P, 10\%, 20)$$

$$= (\$92,946.14)(0.1175)$$

$$= \$10,921$$

Similarly, evaluate alternative B.

$$P_B = \$35,000 + (\$3000)(P/A, 10\%, 5)$$

$$+ (\$4000)(P/A, 10\%, 5)(P/F, 10\%, 5)$$

$$(P/A, 10\%, 5) = 3.7908$$

$$(P/F, 10\%, 5) = 0.6209$$

$$P_B = \$35,000 + (\$3000)(3.7908)$$

$$+ (\$4000)(3.7908)(0.6209)$$

$$= \$55,787.23$$

$$\text{EUAC(B)} = (\$55,787.23)(A/P, 10\%, 10)$$

$$= (\$55,787.23)(0.1627)$$

$$= \$9077$$

Because EUAC(B) is less than EUAC(A),

> Alternative B is economically superior.

23. For both cases, if the annual cost is compared with a total annual mileage of M,

$$A_A = \$0.25M$$

$$A_B = (\$30,000)(A/P, 10\%, 3) + \$0.14M$$
$$+ \$500 - (\$7200)(A/F, 10\%, 3)$$

$$(A/P, 10\%, 3) = 0.4021$$

$$(A/F, 10\%, 3) = 0.3021$$

$$A_B = (\$30,000)(0.4021) + \$0.14M + \$500$$
$$- (\$7200)(0.3021)$$
$$= \$12,063 + \$0.14M$$
$$+ \$500 - \$2175.12$$
$$= \$10,387.88 + \$0.14M$$

For an equal annual cost $A_A = A_B$,

$$\$0.25M = \$10,387.88 + \$0.14M$$

An annual mileage would be $M = 94,435$ mi.

For an annual mileage less than that, $A_A < A_B$.

> Plan A is economically superior until that mileage is exceeded.

24. (a) For method A,

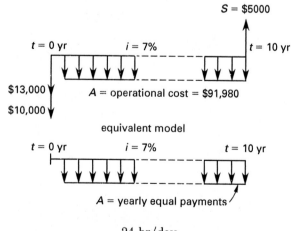

24 hr/day
365 days/yr
total of $(24)(365) = 8760$ hr/yr
$10.50 operational cost/hr
total of $(8760)(\$10.50) = \$91,980$ operational cost/yr

$$A = \$91,980 + (\$23,000)(A/P, 7\%, 10)$$
$$- (\$5000)(A/F, 7\%, 10)$$

$$(A/P, 7\%, 10) = 0.1424$$

$$(A/F, 7\%, 10) = 0.0724$$

$$A = \$91,980 + (\$23,000)(0.1424)$$
$$- (\$5000)(0.0724)$$
$$= \$94,893.30/\text{yr}$$

Therefore, the uniform annual cost per ton each year is

$$\frac{\$94,893.30}{50 \text{ ton}} = \boxed{\$1897.87}$$

For method B,

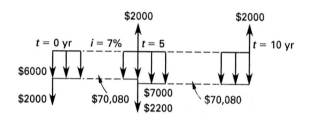

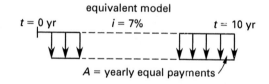

8760 hr/yr
$8 operational cost/hr
total of $(8760)(\$8) = \$70,080$ operational cost/yr

$$A = \$70,080$$
$$+ (\$6000 + \$2000)(A/P, 7\%, 10)$$
$$+ (\$7000 + \$2200 - \$2000)$$
$$\times (P/F, 7\%, 5)$$
$$\times (A/P, 7\%, 10)$$
$$- (\$2000)(A/F, 7\%, 10)$$

$$(A/P, 7\%, 10) = 0.1424$$

$$(A/F, 7\%, 10) = 0.07238$$

$$(P/F, 7\%, 5) = 0.7130$$

$$A = \$70,080 + (\$8000)(0.1424)$$
$$+ (\$7200)(0.7130)(0.1424)$$
$$- (\$2000)(0.0724)$$
$$= \$71,805.42/\text{yr}$$

Therefore, the uniform annual cost per ton each year is

$$\frac{\$71,805.42}{20 \text{ ton}} = \boxed{\$3590.27}$$

(b)

tons/yr	cost of using A (\$)		cost of using B (\$)		cheapest
0–20	94,893	(1x)	71,805	(1x)	B
20–40	94,893	(1x)	143,610	(2x)	A
40–50	94,893	(1x)	215,415	(3x)	A
50–60	189,786	(2x)	215,415	(3x)	A
60–80	189,786	(2x)	287,220	(4x)	A

25.

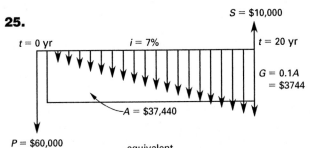

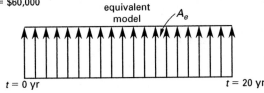

$$A_e = (\$60,000)(A/P, 7\%, 20) + A$$
$$+ G(P/G, 7\%, 20)(A/P, 7\%, 20)$$
$$- (\$10,000)(A/F, 7\%, 20)$$

$$(A/P, 7\%, 20) = 0.0944$$

$$A = (37,440 \text{ mi})\left(1.0 \frac{\$}{\text{mi}}\right) = \$37,440$$

$$G = 0.1A = (0.1)(\$37,440) = \$3744$$

$$(P/G, 7\%, 20) = 77.5091$$

$$(A/F, 7\%, 20) = 0.0244$$

$$A_e = (\$60,000)(0.0944) + \$37,440$$
$$+ (\$3744)(77.509)(0.0944)$$
$$- (\$10,000)(0.0244)$$
$$= \$70,253.78$$

(a) With 80,000 passengers per year, the break-even fare per passenger would be

$$\text{fare} = \frac{A_e}{80,000} = \frac{\$70,253.78}{80,000}$$

$$= \boxed{\$0.878/\text{passenger}}$$

(b) $\qquad \$0.878 = \$0.35 + G(A/G, 7\%, 20)$

$$G = \frac{\$0.878 - \$0.35}{7.3163}$$

$$= \boxed{\$0.072 \text{ increase per year}}$$

(c) As in part (b), the subsidy should be

subsidy = cost − revenue

$$P = \$0.878 - \big(\$0.35 + (\$0.05)(A/G, 7\%, 20)\big)$$
$$= \$0.878 - \big(\$0.35 + (\$0.05)(7.3163)\big)$$
$$= \boxed{\$0.162}$$

26. $\qquad P(A) = (\$4800)(P/A, 12\%, 25)$
$$= (\$4800)(7.8431)$$
$$= \$37,646.88$$

(4 quarters)(25 yr) = 100 compounding periods

$$P(B) = (\$1200)(P/A, 3\%, 100)$$
$$= (\$1200)(31.5989)$$
$$= \$37,918.68$$

$\boxed{\text{Alternative B is economically superior.}}$

27. $\quad$ EUAC(A) = (\$1500)(A/P, 15\%, 5) + \$800
$$= (\$1500)(0.2983) + \$800$$
$$= \$1247.45$$

EUAC(B) = (\$2500)(A/P, 15\%, 8) + \$650
$$= (\$2500)(0.2229) + \$650$$
$$= \$1207.25$$

$\boxed{\text{Alternative B is economically superior.}}$

28. The given data imply that both investments return 4% or more. However, the increased investment of \$30,000 may not be cost effective. Do an incremental analysis.

incremental cost = \$70,000 − \$40,000 = \$30,000

incremental income = \$5620 − \$4075 = \$1545
$$0 = -\$30,000$$
$$+ (\$1545)(P/A, i\%, 20)$$
$$(P/A, i\%, 20) = 19.417$$
$$i \approx 0.25\% < 4\%$$

> Alternative B is economically superior.

(The same conclusion could be reached by taking the present worths of both alternatives at 4%.)

29. (a) $P(\text{A}) = (-\$3000)(P/A, 25\%, 25) - \$20,000$
$$= (-\$3000)(3.9849) - \$20,000$$
$$= -\$31,954.70$$
$$P(\text{B}) = (-\$2500)(3.9849) - \$25,000$$
$$= -\$34,962.25$$

> Alternative A is better.

(b) $\quad \text{CC(A)} = \$20,000 + \dfrac{\$3000}{0.25} = \$32,000$

$\quad \text{CC(B)} = \$25,000 + \dfrac{\$2500}{0.25} = \$35,000$

> Alternative A is better.

(c) $\text{EUAC(A)} = (\$20,000)(A/P, 25\%, 25) + \$3000$
$$= (\$20,000)(0.2509) + \$3000$$
$$= \$8018.00$$
$$\text{EUAC(B)} = (\$25,000)(0.2509) + \$2500$$
$$= \$8772.50$$

> Alternative A is better.

30. (a) $\quad \text{BV} = \$18,000 - (5)\left(\dfrac{\$18,000 - \$2000}{8}\right)$

$\quad = \boxed{\$8000}$

(b) With the sinking fund method, the basis is

$\quad (\$18,000 - \$2000)(A/F, 8\%, 8)$
$$= (\$18,000 - \$2000)(0.0940)$$
$$= \$1504$$

$\quad D_1 = (\$1504)(1.000)$

$\quad = \boxed{\$1504}$

$\quad D_2 = (\$1504)(1.0800)$

$\quad = \boxed{\$1624}$

$\quad D_3 = (\$1504)(1.0800)^2 = (\$1504)(1.1664)$

$\quad = \boxed{\$1754}$

(c) $\quad D_1 = \left(\tfrac{2}{8}\right)(\$18,000)$

$\quad = \boxed{\$4500}$

$\quad D_2 = \left(\tfrac{2}{8}\right)(\$18,000 - \$4500)$

$\quad = \boxed{\$3375}$

$\quad D_3 = \left(\tfrac{2}{8}\right)(\$18,000 - \$4500 - \$3375)$

$\quad = \boxed{\$2531}$

$\quad D_4 = \left(\tfrac{2}{8}\right)(\$18,000 - \$4500 - \$3375 - \$2531)$

$\quad = \boxed{\$1898}$

$\quad D_5 = \left(\tfrac{2}{8}\right)(\$18,000 - \$4500 - \$3375$
$\qquad\qquad - \$2531 - \$1898)$

$\quad = \boxed{\$1424}$

$\quad \text{BV} = \$18,000 - \$4500 - \$3375 - \$2531$
$\qquad\qquad - \$1898 - \$1424$

$\quad = \boxed{\$4272}$

31. $\quad T = \left(\tfrac{1}{2}\right)(8)(9) = 36$

$\quad D_1 = \left(\tfrac{8}{36}\right)(\$14,000 - \$1800) = \$2711$

$\quad \Delta D = \left(\tfrac{1}{36}\right)(\$14,000 - \$1800) = \$339$

$\quad D_2 = \$2711 - \$339 = \$2372$

$\quad \text{DR} = (0.52)(\$2711)(P/A, 15\%, 8)$
$\qquad\quad - (0.52)(\$339)(P/G, 15\%, 8)$
$$= (0.52)(\$2711)(4.4873)$$
$\qquad\quad - (0.52)(\$339)(12.4807)$

$\quad = \boxed{\$4125.74}$

32. $\text{EUAC}_{5\text{ ft}} = (\$600,000)(A/P, 12\%, 15)$
$\qquad\qquad + \left(\tfrac{14}{50}\right)(\$600,000)$
$\qquad\qquad + \left(\tfrac{8}{50}\right)(\$650,000)$
$\qquad\qquad + \left(\tfrac{3}{50}\right)(\$700,000)$
$\qquad\qquad + \left(\tfrac{1}{50}\right)(\$800,000)$
$\qquad\quad = \$418,080$

$$\text{EUAC}_{10 \text{ ft}} = (\$710{,}000)(A/P, 12\%, 15)$$
$$+ \left(\tfrac{8}{50}\right)(\$650{,}000)$$
$$+ \left(\tfrac{3}{50}\right)(\$700{,}000)$$
$$+ \left(\tfrac{1}{50}\right)(\$800{,}000)$$
$$= \$266{,}228$$

$$\text{EUAC}_{15 \text{ ft}} = (\$900{,}000)(A/P, 12\%, 15)$$
$$+ \left(\tfrac{3}{50}\right)(\$700{,}000)$$
$$+ \left(\tfrac{1}{50}\right)(\$800{,}000)$$
$$= \$190{,}120$$

$$\text{EUAC}_{20 \text{ ft}} = (\$1{,}000{,}000)(A/P, 12\%, 15)$$
$$+ \left(\tfrac{1}{50}\right)(\$800{,}000)$$
$$= \$162{,}800$$

Build to 20 ft.

33. Assume replacement after one year.
$$\text{EUAC}(1) = (\$30{,}000)(A/P, 12\%, 1) + \$18{,}700$$
$$= (\$30{,}000)(1.12) + \$18{,}700$$
$$= \$52{,}300$$

Assume replacement after two years.
$$\text{EUAC}(2) = (\$30{,}000)(A/P, 12\%, 2)$$
$$+ \$18{,}700 + (\$1200)(A/G, 12\%, 2)$$
$$= (\$30{,}000)(0.5917) + \$18{,}700$$
$$+ (\$1200)(0.4717)$$
$$= \$37{,}017$$

Assume replacement after three years.
$$\text{EUAC}(3) = (\$30{,}000)(A/P, 12\%, 3)$$
$$+ \$18{,}700 + (\$1200)(A/G, 12\%, 3)$$
$$= (\$30{,}000)(0.4163) + \$18{,}700$$
$$+ (\$1200)(0.9246)$$
$$= \$32{,}299$$

Similarly, calculate to obtain the numbers in the following table.

years in service	EUAC ($)
1	52,300
2	37,017
3	32,299
4	30,207
5	29,152
6	28,602
7	28,335
8	28,234
9	28,240
10	28,312

Replace after 8 yr.

34. Assume the head and horsepower data are already reflected in the hourly operating costs.

Let N be the number of hours operated each year.

$$\text{EUAC}(A) = (\$3600 + \$3050)(A/P, 12\%, 10)$$
$$- (\$200)(A/F, 12\%, 10) + 0.30N$$
$$= (\$6650)(0.1770)$$
$$- (\$200)(0.0570) + 0.30N$$
$$= 1165.65 + 0.30N$$

$$\text{EUAC}(B) = (\$2800 + \$5010)(A/P, 12\%, 10)$$
$$- (\$280)(A/F, 12\%, 10) + 0.10N$$
$$= (\$7810)(0.1770)$$
$$- (\$280)(0.0570) + 0.10N$$
$$= 1366.41 + 0.10N$$

$$\text{EUAC}(A) = \text{EUAC}(B)$$
$$1165.65 + 0.30N = 1366.41 + 0.10N$$

$$N = \boxed{1003.8 \text{ hr}}$$

35. (a) From Eq. 68.78,

$$\frac{T_2}{T_1} = 0.88 = 2^{-b}$$

$$\log 0.88 = -b \log 2$$

$$-0.0555 = -(0.3010)b$$

$$b = 0.1843$$

$$T_4 = (6)(4)^{-0.1843}$$

$$= \boxed{4.65 \text{ wk}}$$

(b) From Eq. 68.79,

$$T_{6\text{--}14} = \left(\frac{6}{1 - 0.1843}\right)$$
$$\times \left(\left(14 + \tfrac{1}{2}\right)^{1-0.1843} - \left(6 - \tfrac{1}{2}\right)^{1-0.1843}\right)$$
$$= \left(\frac{6}{0.8157}\right)(8.857 - 4.017)$$
$$= \boxed{35.6 \text{ wk}}$$

36. First check that both alternatives have an ROR greater than the MARR. Work in thousands of dollars.

Evaluate alternative A.

$$P(A) = -\$120 + (\$15)(P/F, i\%, 5)$$
$$+ (\$57)(P/A, i\%, 5)(1 - 0.45)$$
$$+ \left(\frac{\$120 - \$15}{5}\right)(P/A, i\%, 5)(0.45)$$
$$= -\$120 + (\$15)(P/F, i\%, 5)$$
$$+ (\$40.8)(P/A, i\%, 5)$$

Try 15%.

$$P(A) = -\$120 + (\$15)(0.4972) + (\$40.8)(3.3522)$$
$$= \$24.23$$

Try 25%.

$$P(A) = -\$120 + (\$15)(0.3277) + (\$40.8)(2.6893)$$
$$= -\$5.36$$

Because $P(A)$ goes through zero,

$$(\text{ROR})_A > \text{MARR} = 15\%$$

Next, evaluate alternative B.

$$P(B) = -\$170 + (\$20)(P/F, i\%, 5)$$
$$+ (\$67)(P/A, i\%, 5)(1 - 0.45)$$
$$+ \left(\frac{\$170 - \$20}{5}\right)(P/A, i\%, 5)(0.45)$$
$$= -\$170 + (\$20)(P/F, i\%, 5)$$
$$+ (\$50.35)(P/A, i\%, 5)$$

Try 15%.

$$P(B) = -\$170 + (\$20)(0.4972) + (\$50.35)(3.352)$$
$$= \$8.72$$

Because $P(B) > 0$ and will decrease as i increases,

$$(\text{ROR})_B > 15\%$$

ROR > MARR for both alternatives.

Do an incremental analysis to see if it is worthwhile to invest the extra $170 - \$120 = \$50$.

$$P(B - A) = -\$50 + (\$20 - \$15)(P/F, i\%, 5)$$
$$+ (\$50.35 - \$40.8)(P/A, i\%, 5)$$

Try 15%.

$$P(B - A) = -\$50 + (\$5)(0.4972)$$
$$+ (\$9.55)(3.3522)$$
$$= -\$15.50$$

Because $P(B - A)$ is less than 0 and would become more negative as i increased, the ROR of the added investment is less than 15%.

$$\boxed{\text{Alternative A is superior.}}$$

37. Use the year-end convention with the tax credit. The purchase is made at $t = 0$ yr. However, the tax credit is received at $t = 1$ yr and must be multiplied by $(P/F, i\%, 1)$.

(Note that 0.0667 is actually two-thirds of 10%.)

$$P = -\$300,000 + (0.0667)(\$300,000)(P/F, i\%, 1)$$
$$+ (\$90,000)(P/A, i\%, 5)(1 - 0.48)$$
$$+ \left(\frac{\$300,000 - \$50,000}{5}\right)(P/A, i\%, 5)(0.48)$$
$$+ (\$50,000)(P/F, i\%, 5)$$
$$= -\$300,000 + (\$20,000)(P/F, i\%, 1)$$
$$+ (\$46,800)(P/A, i\%, 5)$$
$$+ (\$24,000)(P/A, i\%, 5)$$
$$+ (\$50,000)(P/F, i\%, 5)$$

By trial and error,

i (%)	P ($\$$)
10	17,616
15	−20,412
12	1448
13	− 6142
$12\frac{1}{4}$	− 479

$$\boxed{i \text{ is between 12 and } 12^{1}/_{4}\%.}$$

38. Assume loan payments are made at the end of each year. Find the annual payment.

$$\text{payment} = (\$2,500,000)(A/P, 12\%, 25)$$
$$= (\$2,500,000)(0.1275)$$
$$= \$318,750$$

$$\text{distributed profit} = (0.15)(\$2,500,000)$$
$$= \$375,000$$

After paying all expenses and distributing the 15% profit, the remainder should be zero.

$$0 = \text{EUAC} = \$20{,}000 + \$50{,}000 + \$200{,}000$$
$$+ \$375{,}000 + \$318{,}750$$
$$- \text{annual receipts}$$
$$- (\$500{,}000)(A/F, 15\%, 25)$$
$$= \$963{,}750 - \text{annual receipts}$$
$$- (\$500{,}000)(0.0047)$$

This calculation assumes $i = 15\%$, which equals the desired return. However, this assumption only affects the salvage calculation, and because the number is so small, the analysis is not sensitive to the assumption.

$$\text{annual receipts} = \$961{,}400$$

The average daily receipts are

$$\frac{\$961{,}400}{365} = \$2634$$

Use the expected value approach. The average occupancy is

$$(0.40)(0.65) + (0.30)(0.70)$$
$$+ (0.20)(0.75) + (0.10)(0.80) = 0.70$$

The average number of rooms occupied each night is

$$(0.70)(120 \text{ rooms}) = 84 \text{ rooms}$$

The minimum required average daily rate per room is

$$\frac{\$2634}{84} = \boxed{\$31.36}$$

39. $\quad \dfrac{\text{annual}}{\text{savings}} = \left(\dfrac{0.69 - 0.47}{1000}\right)(\$3{,}500{,}000) = \$770$

$$P = -\$7500 + (\$770 - \$200 - \$100)$$
$$\times (P/A, i\%, 25)$$
$$= 0$$
$$(P/A, i\%, 25) = 15.957$$

Searching the tables and interpolating,

$$i \approx \boxed{3.75\%}$$

40. Work in millions of dollars.

$$P(A) = -(\$1.3)(1 - 0.48)(P/A, 15\%, 6)$$
$$= (\$1.3)(0.52)(3.7845)$$
$$= -\$2.56 \quad [\text{millions}]$$

Because this is an after-tax analysis and the salvage value is mentioned, assume that the improvements can be depreciated.

Use straight line depreciation.

Evaluate alternative B.

$$D_j = \frac{2}{6} = 0.333$$
$$P(B) = -\$2 - (0.20)(\$1.3)(1 - 0.48)(P/A, 15\%, 6)$$
$$- (\$0.15)(1 - 0.48)(P/A, 15\%, 6)$$
$$+ (\$0.333)(0.48)(P/A, 15\%, 6)$$
$$= -\$2 - (0.20)(\$1.3)(0.52)(3.7845)$$
$$- (\$0.15)(0.52)(3.7845)$$
$$+ (\$0.333)(0.48)(3.7845)$$
$$= -\$2.206 \quad [\text{millions}]$$

Next, evaluate alternative C.

$$D_j = \frac{1.2}{3} = 0.4$$
$$P(C) = -(\$1.2)\big(1 + (P/F, 15\%, 3)\big)$$
$$- (0.20)(\$1.3)(1 - 0.48)$$
$$\times \big((P/F, 15\%, 1) + (P/F, 15\%, 4)\big)$$
$$- (0.45)(\$1.3)(1 - 0.48)$$
$$\times \big((P/F, 15\%, 2) + (P/F, 15\%, 5)\big)$$
$$- (0.80)(\$1.3)(1 - 0.48)$$
$$\times \big((P/F, 15\%, 3) + (P/F, 15\%, 6)\big)$$
$$+ (\$0.4)(0.48)(P/A, 15\%, 6)$$
$$= -(\$1.2)(1.6575)$$
$$- (0.20)(\$1.3)(0.52)(0.8696 + 0.5718)$$
$$- (0.45)(\$1.3)(0.52)(0.7561 + 0.4972)$$
$$- (0.80)(\$1.3)(0.52)(0.6575 + 0.4323)$$
$$+ (\$0.4)(0.48)(3.7845)$$
$$= -\$2.436 \quad [\text{millions}]$$

$$\boxed{\text{Alternative B is superior.}}$$

41. This is a replacement study. Because production capacity and efficiency are not a problem with the defender, the only question is when to bring in the challenger.

Because this is a before-tax problem, depreciation is not a factor, nor is book value.

The cost of keeping the defender one more year is

$$\text{EUAC(defender)} = \$200,000 + (0.15)(\$400,000)$$
$$= \$260,000$$

For the challenger,

$$\text{EUAC(challenger)}$$
$$= (\$800,000)(A/P, 15\%, 10) + \$40,000$$
$$+ (\$30,000)(A/G, 15\%, 10)$$
$$- (\$100,000)(A/F, 15\%, 10)$$
$$= (\$800,000)(0.1993) + \$40,000$$
$$+ (\$30,000)(3.3832)$$
$$- (\$100,000)(0.0493)$$
$$= \$296,006$$

Keep the defender because it is cheaper. The same analysis next year will give identical answers. Therefore, keep the defender for the next three years, at which time the decision to buy the challenger will be automatic.

Having determined that it is less expensive to keep the defender than to maintain the challenger for ten years, determine whether the challenger is less expensive if retired before ten years.

If retired in nine years,

$$\text{EUAC(challenger)} = (\$800,000)(A/P, 15\%, 9)$$
$$+ \$40,000$$
$$+ (\$30,000)(A/G, 15\%, 9)$$
$$- (\$150,000)(A/F, 15\%, 9)$$
$$= (\$800,000)(0.2096)$$
$$+ \$40,000 + (\$30,000)(3.0922)$$
$$- (\$150,000)(0.0596)$$
$$= \$291,506$$

Similar calculations yield the following results for all the retirement dates.

n	EUAC ($)
10	296,000
9	291,506
8	287,179
7	283,214
6	280,016
5	278,419
4	279,909
3	288,013
2	313,483
1	360,000

Because none of these equivalent uniform annual costs is less than that of the defender, it is not economical to buy and keep the challenger for any length of time.

Keep the defender.

69 Engineering Law

PRACTICE PROBLEMS

1. List the different forms of company ownership. What are the advantages and disadvantages of each?

2. Define the requirements for a contract to be enforceable.

3. What standard features should a written contract include?

4. Describe the ways a consulting fee can be structured.

5. What is a retainer fee?

SOLUTIONS

1. The three forms of company ownership are the (1) sole proprietorship, (2) partnership, and (3) corporation.

A *sole proprietor* is his or her own boss. This satisfies the proprietor's ego and facilitates quick decisions. However, unless the proprietor is trained in business, the company will usually operate without the benefit of expert or mitigating advice. The sole proprietor also personally assumes all the debts and liabilities of the company. A sole proprietorship is terminated upon the death of the proprietor.

A *partnership* increases the capitalization and the knowledge base beyond that of a proprietorship, but offers little else in the way of improvement. In fact, the partnership creates an additional disadvantage of one partner's possible irresponsible actions creating debts and liabilities for the remaining partners.

A *corporation* has sizable capitalization (provided by the stockholders) and a vast knowledge base (provided by the board of directors). It keeps the company's and the owner's liabilities separate. It also survives the death of any employee, officer, or director. Its major disadvantage is the administrative work required to establish and maintain the corporate structure.

2. To be legal, a contract must contain (1) an offer, (2) some form of consideration (which does not have to be equitable), and (3) an acceptance by both parties. To be enforceable, the contract must be voluntarily entered into, both parties must be competent and of legal age, and the contract cannot be for illegal activities.

3. A written contract will identify both parties, state the purpose of the contract and the obligations of the parties, give specific details of the obligations (including relevant dates and deadlines), specify the consideration, state the boilerplate clauses to clarify the contract terms, and leave places for signatures.

4. A consultant will either charge a fixed fee, a variable fee, or some combination of the two. A one-time fixed fee is known as a lump-sum fee. In a cost plus fixed fee contract, the consultant will also pass on certain costs to the client. Some charges to the client may depend on other factors, such as the salary of the consultant's staff, the number of days the consultant works, or the eventual cost or value of an item being designed by the consultant.

5. A *retainer* is a (usually) nonreturnable advance paid by the client to the consultant. While the retainer may be intended to cover the consultant's initial expenses until the first big billing is sent out, there does not need to be any rational basis for the retainer. Often, a small retainer is used by the consultant to qualify the client (to ensure the client is not just shopping around and getting free initial consultations) and as a security deposit (to ensure the client does not change consultants after work begins).

70 Engineering Ethics

PRACTICE PROBLEMS

(Note: Each problem has two parts. Determine whether the situation is (or can be) permitted legally. Then, determine whether the situation is permitted ethically.)

1. (a) Is it legal and/or ethical for an engineer to sign and seal plans that were not prepared by him or prepared under his responsible direction, supervision, or control?

(b) Is it legal and/or ethical for an engineer to sign and seal plans that were not prepared by him but were prepared under his responsible direction, supervision, and control?

2. Under what conditions would it be legal and/or ethical for an engineer to rely on the information (e.g., elevations and amounts of cuts and fills) furnished by a grading contractor?

3. Is it legal and/or ethical for an engineer to alter the soils report prepared by another engineer for his client?

4. Under what conditions would it be legal and/or ethical for an engineer to assign work called for in his contract to another engineer?

5. A licensed professional engineer was convicted of a felony totally unrelated to his consulting engineering practice.

(a) What actions should be taken by the state registration board?

(b) What actions should be taken by the professional or technical society (e.g., ASCE, ASME, IEEE, NSPE)?

6. An engineer comes across some work of a predecessor. After verifying the validity and correctness of all assumptions and calculations, the engineer uses the work. Under what conditions would such use be legal and/or ethical?

7. A building contractor makes it a policy to provide cellular car telephones to the engineers of projects he is working on. Under what conditions could the engineers accept the telephones?

8. An engineer designs a tilt-up slab building for a client. The design engineer sends the design out to another engineer for checking. The checking engineer himself sends the plans to a concrete contractor for review. The concrete contractor makes suggestions that are incorporated into the recommendations of the checking contractor. These recommendations are subsequently incorporated into the plans by the original design engineer. What steps must be taken to keep the design process legal and/or ethical?

9. A consulting engineer registers his corporation as "John Williams, P.E. and Associates, Inc." even though he has no associates. Under what conditions would this name be legal and/or ethical?

10. Upon learning that a chemical plant is planning on producing a toxic product, an engineer writes to the local newspaper, condemning the action. Under what conditions would this action be legal and/or ethical?

11. An engineer signs a contract with a client. The fee the client agreed to pay was based on the engineer's estimate of time required. The engineer is able to complete the contract satisfactorily in half the time he expected. Under what conditions would it be legal and/or ethical for the engineer to keep the full fee?

12. After working on a project for a client, an engineer receives an offer to perform design services for that client's competitor. Under what conditions would it be legal and/or ethical for the engineer to work for the competitor?

13. Two engineers submit bids to a prospective client for a design project. The client tells engineer A how much engineer B bid and invites engineer A to beat the amount. Under what conditions could engineer A legally/ethically submit a lower bid?

14. A registered civil engineer specializing in well-drilling, irrigation pipelines, and farmhouse sanitary systems takes a booth at a county fair located in a farming town. By a random drawing, the engineer's booth is located next to a hog-breeder's booth, complete with live (prize) hogs. The engineer gives away helium balloons with her name and phone number to all visitors to the booth. Did the engineer violate any laws/ethical guidelines?

15. While in a developing country supervising construction of a project he designed, an engineer discovers his client's project manager is treating local workers in an unsafe and inhumane (but for that country, legal) manner. When the engineer objects, the client tells him to mind his own business. Later, the local workers ask the engineer to participate in a walkout and strike with them.

(a) What legal/ethical positions should the engineer take?

(b) Should it have made any difference if the engineer had or had not yet accepted any money from the client?

16. While working for a client, an engineer learns confidential knowledge of a proprietary production process being used by the client's chemical plant. The process is clearly destructive to the environment, but the client will not listen to the engineer's objections. Informing the proper authorities would require the engineer to release information that she gained in confidence. Is it legal and/or ethical for the engineer to expose the client?

17. While working for an engineering design firm, an engineer moonlights as a soils engineer. At night, the engineer uses his employer's facilities to analyze and plot the results of soils tests. He then uses his employer's computers and word processors to write his reports. The equipment, computers, and word processors would otherwise be unused. Under what conditions could the engineer's actions be considered legal and/or ethical?

SOLUTIONS

Introduction to the Answers

Case studies in law and ethics can be interpreted in many ways. The problems presented are simple thumbnail outlines. In most real cases, there will be more facts to influence a determination than are presented in the case scenarios. In some cases, a state may have specific laws affecting the determination; in other cases, prior case law will have been established.

The determination of whether an action is legal can be made in two ways. The obvious interpretation of an illegal action is one that violates a specific law or statute. An action can also be *found to be illegal* if it is judged in court to be a breach of a written, verbal, or implied contract. Both of these approaches are used in the following solutions.

These answers have been developed to teach legal and ethical principals. While being realistic, they are not necessarily based on actual incidents or prior case law.

1. (a) Stamping plans for someone else is illegal. The registration laws of all states permit a registered engineer to stamp/sign/seal only plans that were prepared by him personally or were prepared under his direction, supervision, or control. This is sometimes called being in *responsible charge*. The stamping/signing/sealing, for a fee or gratis, of plans produced by another person, whether that person is registered or not and whether that person is an engineer or not, is illegal.

(b) The act is unethical. An illegal act, being a concealed act, is intrinsically unethical. In addition, stamping/signing/sealing plans that have not been checked violates the rule contained in all ethical codes that requires an engineer to protect the public.

2. Unless the engineer and contractor worked together such that the engineer had personal knowledge that the information was correct, accepting the contractor's information is illegal. Not only would using unverified data violate the state's registration law (for the same reason that stamping/signing/sealing unverified plans in problem 1 was illegal), but the engineer's contract clause dealing with assignment of work to others would probably be violated.

The act is unethical. An illegal act, being a concealed act, is intrinsically unethical. In addition, using unverified data violates the rule contained in all ethical codes that requires an engineer to protect the client.

3. It is illegal to alter a report to bring it "more into line" with what the client wants unless the alterations represent actual, verified changed conditions. Even when the alterations are warranted, however, use of the unverified remainder of the report is a violation of the state registration law requiring an engineer only to stamp/sign/seal plans developed by or under him. Furthermore, this would be a case of fraudulent misrepresentation unless the originating engineer's name was removed from the report.

Unless the engineer who wrote the original report has given permission for the modification, altering the report would be unethical.

4. Assignment of engineering work is legal if (1) the engineer's contract permitted assignment, (2) all prerequisites (i.e., notifying the client) were met, and (3) the work was performed under the direction of another licensed engineer.

Assignment of work is ethical if (1) it is not illegal, (2) it is done with the awareness of the client, and (3) the assignor has determined that the assignee is competent in the area of the assignment.

5. (a) The registration laws of many states require a hearing to be held when a licensee is found guilty of unrelated, but nevertheless unforgivable, felonies (e.g., moral turpitude). The specific action (e.g., suspension, revocation of license, public censure, etc.) taken depends on the customs of the state's registration board.

(b) By convention, it is not the responsibility of technical and professional organizations to monitor or judge the personal actions of their members. Such organizations do not have the authority to discipline members (other than to revoke membership), nor are they immune from possible retaliatory libel/slander lawsuits if they publicly censure a member.

6. The action is legal because, by verifying all the assumptions and checking all the calculations, the engineer effectively does the work himself. Very few engineering procedures are truly original; the fact that someone else's effort guided the analysis does not make the action illegal.

The action is probably ethical, particularly if the client and the predecessor are aware of what has happened (although it is not necessary for the predecessor to be told). It is unclear to what extent (if at all) the predecessor should be credited. There could be other extenuating circumstances that would make referring to the original work unethical.

7. Gifts, per se, are not illegal. Unless accepting the phones violates some public policy or other law, or is in some way an illegal bribe to induce the engineer to favor the contractor, it is probably legal to accept the phones.

Ethical acceptance of the phones requires (among other considerations) that (1) the phones be required for the job, (2) the phones be used for business only, (3) the phones are returned to the contractor at the end of the

job, and (4) the contractor's and engineer's clients know and approve of the transaction.

8. There are two issues here: (1) the assignment and (2) the incorporation of work done by another. To avoid a breach, the contracts of both the design and checking engineers must permit the assignments. To avoid a violation of the state registration law requiring engineers to be in responsible charge of the work they stamp/sign/seal, both the design and checking engineers must verify the validity of the changes.

To be ethical, the actions must be legal and all parties (including the design engineer's client) must be aware that the assignments have occurred and that the changes have been made.

9. The name is probably legal. If the name was accepted by the state's corporation registrar, it is a legally formatted name. However, some states have engineering registration laws that restrict what an engineering corporation may be named. For example, all individuals listed in the name (e.g., "Cooper, Williams, and Somerset—Consulting Engineers") may need to be registered. Whether having "Associates" in the name is legal depends on the state.

Using the name is unethical. It misleads the public and represents unfair competition with other engineers running one-person offices.

10. Unless the engineeer's accusation is known to be false or exaggerated, or the engineer has signed an agreement (e.g., confidentiality, non-disclosure, etc.) with his employer forbidding the disclosure, the letter to the newspaper is probably not illegal.

The action is probably unethical. (If the letter to the newspaper is unsigned, it is a concealed action and is definitely unethical.) While whistle-blowing to protect the public is implicitly an ethical procedure, unless the engineer is reasonably certain that manufacture of the toxic product represents a hazard to the public, he has a responsibility to the employer. Even then, the engineer should exhaust all possible remedies to render the manufacture nonhazardous before blowing the whistle. Of course, the engineer may quit working for the chemical plant and be as critical as the law allows without violating engineer-employer ethical considerations.

11. Unless the engineer's payment was explicitly linked in the contract to the amount of time spent on the job, taking the full fee would not be illegal or a breach of the contract.

An engineer has an obligation to be fair in estimates of cost, particularly when the engineer knows no one else is providing a competitive bid. Taking the full fee would be ethical if the original estimate was arrived at logically and was not meant to deceive or take advantage of the client. An engineer is permitted to take advantage of economies of scale, state-of-the-art techniques, and break-through methods. (Similarly, when a job costs more than the estimate, the engineer may be ethically bound to stick with the original estimate.)

12. In the absence of a nondisclosure or noncompetition agreement or similar contract clause, working for the competitor is probably legal.

Working for both clients is unethical. Even if both clients know and approve, it is difficult for the engineer not to "cross-pollinate" his work and improve one client's position with knowledge and insights gained at the expense of the other client. Furthermore, the mere appearance of a conflict of interest of this type is a violation of most ethical codes.

13. In the absence of a sealed-bid provision mandated by a public agency and requiring all bids to be opened at once (and the award going to the lowest bidder), the action is probably legal.

It is unethical for an engineer to undercut the price of another engineer. Not only does this violate a standard of behavior expected of professionals, it unfairly benefits one engineer because a similar chance is not given to the other engineer. Even if both engineers are bidding openly against each other (in an auction format), the client must understand that a lower price means reduced service. Each reduction in price is an incentive to the engineer to reduce the quality or quantity of service.

14. It is generally legal for an engineer to advertise his services. Unless the state has relevant laws, the engineer probably did not engage in illegal actions.

Most ethical codes prohibit unprofessional advertising. The unfortunate location due to a random drawing might be excusable, but the engineer should probably refuse to participate. In any case, the balloons are a form of unprofessional advertising, and as such, are unethical.

15. (a) As stated in the scenario statement, the client's actions are legal for that country. The fact that the actions might be illegal in another country is irrelevant. Whether or not the strike is legal depends on the industry and the laws of the land. Some or all occupations (e.g., police and medical personnel) may be forbidden to strike. Assuming the engineer's contract does not prohibit his own participation, the engineer should determine the legality of the strike before making a decision to participate.

If the client's actions are inhuman, the engineer has an ethical obligation to withdraw from the project. Not doing so associates the profession of engineering with human misery.

(b) The engineer has a contract to complete the project for the client. (It is assumed that the contract between the engineer and client was negotiated in good faith, that the engineer had no knowledge of the work conditions prior to signing, and that the client did not falsely induce the engineer to sign.) Regardless of the reason for withdrawing, the engineer is breaching his contract. In the absence of proof of illegal actions by the client, withdrawal by the engineer requires a return of all fees received. Even if no fees have been received, withdrawal exposes the engineer to other delay-related claims by the client.

16. A contract for an illegal action cannot be enforced. Therefore, any confidentiality or nondisclosure agreement that the engineer has signed is unenforceable if the production process is illegal, uses illegal chemicals, or violates laws protecting the environment. If the production process is not illegal, it is not legal for the engineer to expose the client.

Society and the public are at the top of the hierarchy of an engineer's responsibilities. Obligations to the public take precedence over those to the client. If the production process is illegal, it would be ethical to expose the client.

17. It is probably legal for the engineer to use the facilities, particularly if the employer is aware of the use. (The question of whether the engineer is trespassing or violating a company policy cannot be answered without additional information.)

Moonlighting, in general, is not ethical. Most ethical codes prohibit running an engineering consulting business while receiving a salary from another employer. The rationale is that the moonlighting engineer is able to offer services at a much lower price, placing other consulting engineers at a competitive disadvantage. The use of someone else's equipment only compounds the problem. Because the engineer does not have to pay for using the equipment, he does not have to charge his clients for it. This places him at an unfair competitive advantage compared to other consultants who have invested heavily in equipment.

71 Electrical Engineering Frontiers

PRACTICE PROBLEMS

1. What are the force particles responsible for the electromagnetic interaction?

- (A) bosons
- (B) gluons
- (C) gravitons
- (D) photons

SOLUTIONS

1. The force particles for electromagnetic interactions are photons, sometimes called virtual photons because they cannot be detected without interfering with the interaction.

The answer is (D).

72 Engineering Licensing in the United States

PRACTICE PROBLEMS

1. What might be a reason to obtain a PE license?

(A) increased respect

(B) increased wages

(C) proof of one's engineering knowledge

(D) all of the above

SOLUTIONS

1. Obtaining a PE license is proof to peers of the broad base of your knowledge, and an example of your self discipline. The PE will engender greater respect and is known to have a positive impact on salary. Therefore, D, all of the above, is the answer.